SELF EVIDENT

SELF EVIDENT

*Undeniable Proof that Science is Discovering God
and Possibly Revealing His Plan for Humanity*

DON BONGAARDS P. E.

SELF EVIDENT

Copyright © 2016 by Don Bongaards, P.E.

World Ahead Press is a division of WND Books. The views and opinions expressed in this book are those of the author and do not necessarily reflect the official policy or position or WND Books.

Paperback ISBN: 978-1-944212-56-8
eBook ISBN: 978-1-944212-57-5

Printed in the United States of America
16 17 18 19 20 21 LSI 9 8 7 6 5 4 3 2 1

CONTENTS

DEDICATION

I dedicate this book to my four grandsons, Britt, Stewart, Johnathan, and Evan. I pray they will live long, healthy, prosperous, and happy lives.

I also want to provide a special dedication of this book to GOD as a small token of my appreciation and gratitude for all He has done for me in my life.

BEFORE READING THIS BOOK
CONSIDER THESE QUOTES FROM
ALBERT EINSTEIN

"Science without religion is lame, religion without science is blind."

"The more I study science the more I believe in God."

"When the solution is simple, God is answering."

"I want to know how God created this world. I am not interested in this or that phenomenon in the spectrum of this or that element. I want to know His thoughts; the rest are details."

"There are two ways to live your life. One is though nothing is a miracle. The other is though everything is a miracle."

"No one can read the Gospels without feeling the actual presence of Jesus. His personality pulsates in every word."

CAUTION

READING THIS BOOK MAY CAUSE COGNITIVE DISSONANCE

Cognitive dissonance may occur when scientific data provides ever more convincing arguments against a favored paradigm. It causes all sorts of mental machinations that allow us to retain our preconceived notions of reality. It's humanity's inherent ability to ignore unpleasant facts. This is especially true among scientists who have spent much of a lifetime trying to prove the validity of a premise in question. The emotional stakes are high.

PROLOGUE

The title of this book is *Self Evident*, which is a shortened way of saying, "The logic of what you are about to read should be obvious to any thinking person." Moreover, with the subtitle "Undeniable Proof That Science is Discovering God and Possibly Revealing His Plan for Humanity," the reader should expect to discover some little known scientific facts that unquestionably point directly toward God as the designer and creator of the universe. It also implies that God is using recent scientific facts to tell us why we humans were created and what our purpose might be. So be prepared for what may be a life changing experience.

Before we begin though, I want to take you on an excursion into the future. This is a requirement for you to understand where I'm coming from. It's a requirement because it shows how an engineer sees the evolution of today's technology, and how it could affect our thinking about today's circumstances. For example, if you were alive hundreds of years ago, you probably thought the Earth was flat and the only explanation for your existence was that God made it that way. Today, the idea of a flat Earth is absurd, and because of Charles Darwin's theory of macro-evolution, many people believe that God is a myth. My guess is that people living hundreds of years into the future will see many things that we believe now as absurd.

As an engineer, my interest in the future is more associated with "game-changing" inventions and ideas that will alter the way we think and live, rather than with how governments, laws, and the media can affect the way we think and live.

Let's begin our excursion into the future with the idea that in fifty years the United States becomes prosperous and more influential in world affairs, including providing the energy, water, technology and partial funding needed to bring the world's impoverished people from poverty to prosperity. In fifty years, the world's population will have grown to over ten billion, and nuclear fusion energy will begin to emerge. We might have also developed a Moon base and begun transporting high-valued metals and helium 3 back to Earth. Helium 3 is in itself a game-changer, in that it is readily available in space and is the most appropriate fuel for nuclear fusion energy.

Transporting materials from the Moon to Earth will require an intermediate step. A robotically built orbiting space station will be needed to process the Moon materials in zero gravity. The space station will most likely be a cylinder that rotates to simulate gravity on its inner surface (a NASA proposal made many years ago), and may also, for economic reasons, serve a secondary purpose of transmitting solar-derived electrical energy to Earth in the form of microwaves.

With the above conditions as a starting point, let's go further into the future to see how technology advances, and human lives are changed. For purposes I'll explain later, I won't refer to the Bible, since at this point many might think God's plan for our future is already spelled out and whatever I say about the future is meaningless. To this I say   keep in mind that Christ's return may not occur for another thousand years or more and God's plan may require future technologies to exist first. Humans have thought that Christ's return has been imminent for more than two thousand years; why not a thousand more? Although I do believe in biblical prophecy, what we see happening today may be misleading.

Recent history has shown many of our modern-day inventions were derived from science fiction books and movies. Although Leonardo da Vinci didn't write a book, many illustrations in his sketchbook show his unusual ability to predict future inventions such as weapons of war, flying machines, water systems, and work tools. Likewise, Jules Verne predicted battery-powered submarines, helicopters, and rocket ships. In recent times, the televised *Star Trek* series provided some ideas that may still come to pass, like warp speed, phaser weapons, holodecks, molecular transportation, and tractor beams. While these future predictions may serve to capture our imagination, they are more geared toward entertaining television and movie viewers.

Well, my predictions of the future are considerably different from that of *Star Trek*. I'm not predicting encounters with unusual beings like Klingons, or space battles that endanger planet Earth. In fact, my predicted spaceship varies considerably from the Starship *Enterprise*. And, while my predictions may be less entertaining, I can assure you that they are equally intriguing; perhaps even more so since they are extrapolated from currently known science. To me, this is an important aspect, since it's easier for most people to identify with and understand.

I agree that predictions about the future are probably best described in the form of a science fiction story. So, that's what I'll do. My story is excerpted from a book that I may publish next. It's titled *Epsilon Eridani*, and it's about a multi-generational spacecraft that travels to the Epsilon Eridani solar system to investigate an Earth-like planet. I've gone out of my way to make the book exciting and entertaining, but contained within its pages are many coincidences related to Earth's formation, human evolution, and history; and these coincidences differ considerably from what many people have been led to believe. The following is excerpted from the beginning of Chapter 1, and another chapter that describes some things I think you will find quite fascinating.

THE EPSILON ERIDANI STORY

Hello my name is William Harvath. It's the year 2360. I'm one hundred and thirty-six years old. I live on a very large spacecraft, and I will be your tour guide. We are currently orbiting the Epsilon Eridani planet that contains animals and what appears to be intelligent but primitive life. It has taken one hundred and ten years for our spacecraft to get here from planet Earth, and my great-grandson, Jason Harvath, is currently organizing an exploratory team to investigate what's on the planet's surface.

Ok. You want to know more about our spacecraft. Well, as you can see, I'm sitting on a bench in a park located on the top layer of the inside surface. In case you're wondering why I'm not floating around in zero gravity, it's because our cylindrical spacecraft is rotating so that objects, including me, are held onto the inside surface by centrifugal force. It's like the force you feel when you tie a ball onto the end of a string and spin it around.

Before I describe the internal parts of this spacecraft's design, I need to tell you that it's sixteen miles in diameter and forty-eight miles long. Pretty big, huh? It was made almost entirely from materials obtained from the asteroid belt located between Jupiter and Mars. Our oxygen and nitrogen atmosphere was derived from processing asteroid materials and from Jupiter's moon, Titan. When asteroid materials are processed, a derivative of this process is oxygen, and nitrogen was transported in frozen form from Titan's surface. Our water was obtained from the asteroid Ceres and our helium 3 fuel was derived from Jupiter's upper atmosphere. In fact, since we left Earth, more than ten thousand, sixteen-mile by forty-eight mile satellites, pre-spacecraft, have been robotically built, and Earth's population is migrating to them as soon as they become available. When I tell you about how our spacecraft's living spaces are constructed, you will understand why this migration is happening so rapidly.

From what scientists have determined, we could build as many as twelve million of these huge satellites from the enormous amount of materials available from the asteroids in our solar system. With each self-contained satellite being capable of comfortably and luxuriously housing as many as 2.5 million people, our limit could be thirty trillion inhabitants. And, now that I see the materials available here in the Epsilon Eridani solar system, it appears that human population growth could be endless.

Ok. Now let me describe the internal design of our spacecraft; but to do this you need to come with me to one of our museums where they have a high-definition hologram that describes our spacecraft better than I can with words alone. Would you like to go there in an electric vehicle on our roadway system, or using our maglev train? Personally, if you are not in a hurry, I recommend the electric vehicle since it will give you a better view of our upper level common area. Oh, you agree with me? Ok then, I'll signal one of our electric vehicles to pick us up. Boy, that didn't take long. Please get in and I'll tell the vehicle where we need to go.

The first thing I want you to notice is our blue sky and wispy clouds. It's simulated with a combination of LED lighting, a translucent surface, and water vapor. It's almost a mile above our head, so from our vantage point it looks like it did when I lived on Earth. We simulate night and day on a twenty-four-hour basis and at least once a week we have a preplanned weather occurrence like a rainstorm. In the mountain zone we simulate snow storms that keep the ski trails fresh. Notice the fenced-in area that we're now passing; it's one of our wild animal preserves. Safari trips are available on a daily basis if any of you would like to go on one. Yes, there are numerous uniquely designed golf courses. In fact, we are going to pass by one in a few minutes. Yes, again, we have lakes, rivers, streams, waterfalls, flower gardens, and forested areas. We also have schools, universities, sports stadiums, hospitals, churches, and office buildings. Ok, we've arrived. I'll send the vehicle back to its underground parking space, and retrieve it when we are ready to leave. By the way, that vehicle is powered by a miniature helium 3 fusion electric system. Pretty cool, eh?

Ok, here we are at the spacecraft hologram. As you can see, the thick-walled outer shell of the satellite/spacecraft habitat was designed to protect its inhabitants from such things as meteors, solar flares, and cosmic radiation. Obviously, the inside part of the satellite is immune to Earth's problems of hurricanes, earthquakes, tornados, ultraviolet and cosmic radiation, volcanoes, tidal waves, a rising ocean level, and magnetic field reversal. In addition, with on-board sensors, it's capable of maneuvering away from asteroids, comets, and potentially damaging space debris.

The main thing I want you to notice are the stacked circular structures located beneath the top surface common area. They are stacked twelve high and there are one hundred and five thousand of them. They are eight hundred feet in diameter and one hundred feet high, and each family living on board owns one of them. With more than half of these structures being reserved for food, energy, materials processing, water storage, sewage treatment, manufacturing, and other purposes we can accommodate an eventual population of about 2.5 million.

When we started our journey to the Epsilon Eridani solar system, many of these living space structures were left empty to accommodate population growth. In fact, when my wife and son and I left Earth one hundred and ten years ago we began with a population of only two hundred and fifty thousand. During the trip we've grown to more than six hundred thousand. So we have more than enough space to accommodate a growing population for our one hundred and ten-year trip back to planet Earth. I'm sorry to say that our medical advancements haven't progressed to where I'll live to see that day.

If you think our spacecraft is impressive, wait till you learn about how the living space structures are designed; it will blow your mind. First of all, the interior space has an ultra-high definition hologram at the far end that can project anything from a Pacific Ocean scene to a view of a New Hampshire mountain range. In the foreground trees and plants that are indigenous to the holograph scene are arranged to enhance the perceived reality of the view. Further enhancement for a Pacific Ocean view can be provided by adding salt air, the sound of crashing waves, and seagulls. Best of all the holograph and foreground plantings can be changed occasionally for variety.

Notice the uniquely styled house located at the back end of the living space. Generally speaking, most people prefer an upside-down house where the bedrooms are on the bottom floor and the living area is on top. This of course maximizes the magnificent views. With a one-hundred-foot-high ceiling, LED lighting, and holographic imaging, the simulated sky looks very real. Moreover, variations in weather can be programmed as desired.

Now here is the best part. Each family is given an android. The android is fueled by nuclear fusion and can perform any function that humans are able to do. And—you guessed it— they do everything from cooking the meals to manicuring the garden areas. One interesting and important aspect regarding the androids is their ability to repair and replicate themselves.

I think you can understand why Earth's population has been migrating to these living spaces in droves. With an abundance of helium 3 fuel, construction materials, and robotic/android labor, the living spaces and material accommodations are essentially free. It's a form of utopia, but the humans need to supervise android activities and contribute to society. Obviously they have time for raising children, improving their education, and other things that in the past were limited by having to work for a living.

Now for a history lesson to let you know how we got here. Back in the year 2070, a huge space interferometer telescope array had been deployed by NASA to focus on the Earth's closest star systems that were most likely to have Earth-like planets. This imaginative solar system-wide telescope had the ability to see and spectrally analyze planets as small as our Earth's Moon, at a distance as far away as thirty light years. Although it was hoped that our closest star system, Alpha Centauri, at 4.3 light years away, would yield an Earth-like planet, it had long been theorized that this was very unlikely. The reasoning was that in a binary, or trinary system in the case of Alpha Centauri a gaseous planet like Jupiter could not exist because of the heat generated by the adjacent stars. And, without a Jupiter-sized planet, an Earth-like planet would be unlikely. That's because Jupiter, with its enormous gravity, collects up most of the asteroids that may collide with and devastate an Earth-like planet; a space vacuum cleaner, so to speak.

Looking beyond Alpha Centauri, the telescope array discovered an Earth-like planet in the Epsilon Eridani solar system, 10.8 light years from Earth. Although its sensors could not determine the exact composition of the planet, it did show distinct blue, green, and brown colors, which would be indicative of water, vegetation, and land.

Of particular interest was a Moon surrounding this planet that was almost the same size, distance, and orbital rotation as Earth's Moon. The Moon was of special interest since life could not have evolved on Earth without the Moon, because its gravity acts to stabilize the Earth's tilt axis and cause ocean tides.

Of further interest was the existence of gaseous and solid planets that looked very much like Earth's solar system. Unfortunately, the interferometer telescope could not detect objects the size of comets or asteroids. Since no radio signals were being emitted from the planet, speculation about the possibility of primitive life began to grow. As a result of this finding, NASA was given the green light to launch an unmanned flyby mission, a one-way mission that doesn't slow down when it reaches its destination (a fuel consuming requirement) to evaluate the Earth-like planet in more detail. Because of the required one hundred and ten-year flight time, it was decided not to make the trip longer by exploring closer solar systems like Alpha Centauri, Barnard's star, and Sirius.

In the year 2080 the unmanned flyby spacecraft was finally launched to explore the Epsilon Eridani solar system. Upon arrival, one hundred and ten years later, its images and sensor readings were sent back to Earth. However, although transmitted at the speed of light, by the time the radio signals were received back on Earth, it was the year 2200. But, when they were received, the data was astonishing. The probe's telescopic images confirmed the existence of life. In fact, the real probability of human life was inferred by the existence of manmade objects and what appeared to be the movement of animals and human beings.

Fortunately, because of overpopulation and energy shortages experienced in Earth's twenty-first century, humans had begun migrating to live in orbiting space satellites. By the year 2060, the United States had just completed its first commercial demonstration of helium 3-fueled fusion electric power. When joined together in a nuclear fusion reaction, helium 3 atoms produced two electrically charged protons that can generate electricity directly, without requiring the additional step of converting heat into steam. And best of all, the fusion process was completely safe, and did not produce radioactive waste.

At this point in our history lesson, I want to go back to the year 2070, where NASA had established a base on the Moon and deployed a successful mission to Mars. Because they didn't find life on Mars, taxpayers were unwilling to support future space exploration that didn't have a payoff. Since Mars had an elliptical orbit that resulted in extreme temperatures, and didn't have a protective magnetic field, NASA finally admitted that terra-forming Mars for future human habitation was impractical.

It was at this crucial point in time that President Thomas Mitsufugi announced that he had received approval from Congress to redirect NASA's efforts to that of mining helium 3 on the Moon. He also received an agreement from every electric utility company in the United States to build at least one helium 3 fusion electric power plant within five years, and negotiated with other countries that showed a willingness to participate. Since the Moon was not owned by the United States, an offer of technical assistance was made to any country that would like to develop a mining program. Willingness to have an all-out effort to build helium 3 fusion power plants, was a direct result of prior failed energy policies, and a sense of panic as fossil fuels were becoming much more scarce.

The initial phase of the mining program consisted of building a launch ring that used superconducting magnets. This provided a much lower cost method of putting cargo, rather than humans, into space when compared to NASA's abandoned space shuttle program. The cargo, which could withstand very high acceleration forces, permitted mining on the Moon, including the construction of a Moon-based catapult launch system. In addition, a shuttle vehicle was designed to deliver Moon materials between the Moon and Earth orbits.

Besides helium 3, many other valuable elements were found on the Moon, including platinum, which could be used to coat the electrodes for hydrogen fuel cell vehicles that were being used at that time. As it turned out, the Moon's craters contained these valuable elements near its surface. Because the Moon had only one-sixth of Earth's gravity, the valuable minerals, contained in the asteroids that had impacted the Moon remained near the Moon's surface.

As time went by, an improved mining program was initiated by creating a Moon regolith (Moon soil) processing station in space. This processing station consisted of a one-mile diameter by four miles long rotating cylindrical satellite. The rotation simulated Earth's gravity on its inner surface, just like our spacecraft does today, and provided living quarters for workers. Since solar intensity was eight times stronger in space than it is on Earth, several large solar collectors were used to power the regolith processing system and beam excess energy back to Earth using microwaves. This energy was used to power the original electromagnetic launch ring plus several others as time went on.

Thanks to the work of physicist Brandon Johnson, everyone today has their own low-cost helium 3 fusion modules to supply home electricity and power transportation vehicles. And, with space-based robotic collection and processing systems, helium 3 eventually became inexpensive and readily accessible. Helium 3 became inexpensive because by the twenty-second century it was being extracted (in a frozen form) from the upper atmosphere of Jupiter for transport back to Earth.

Upon successful completion of the first mining space station in 2106, a proposal was made to double its size, using residual Moon regolith and iron by-products. Unlike the first space station, the second station was built entirely by robotics. The only human activity was to provide processed Moon and Earth materials and the robots did the rest. Because of improvements made over the previous years, the second space station was completed in only twenty-five years.

By the year 2220, after successive redoubling in size, a sixteen miles in diameter by forty-eight miles long, orbiting cylinder was completed. And, with the addition of helium

3 nuclear powered rocket engines it was converted into a spacecraft that could take humans on a one hundred and ten-year voyage to explore Epsilon Eridani's Earth-like planet. It was named Epsilon 1 to commemorate the voyage that it was about to take.

To ensure sufficient and reliable electric power and propulsion for the round trip, redundant fusion reactors and an excess of fuel were provided. However, twenty-three more years were needed to fill the fuel tanks with helium 3 from the upper atmosphere of Jupiter and from the water containing asteroid Ceres. Seven additional years were required to board a beginning population of two hundred and fifty thousand. That included me and my family.

* * * * *

I could have stopped at this point in the Epsilon Eridani story, but I wanted to include excerpts from chapter 17 because I think you'll find it to be quite interesting and entertaining. It illustrates the contrast between primitive and advanced technologies, and how people's lives and thinking are influenced. This part of the story begins at a point in time when the Epsilon 1 spacecraft is ready to return to Earth, and a family from the Epsilon planet's inhabitants has agreed to return with them. The Epsilon family is partially comprised of Jason's Harvath's son Stewart, who, during an encounter with the Epsilon people, married a girl named Wanaka and had a baby. Wanaka is the granddaughter of the planet's leader, Homanz. The people who lived on the Epsilon planet were quite primitive, similar to ancient Egypt, and at this time in the story had not been exposed to Epsilon 1's technology. So with this introduction, here we go:

CHAPTER 17

Their return to Epsilon 1 was greeted by a large crowd of cheering people and marching bands. Homanz, Wanaka and others in the family are surprised and pleased by this welcoming. The warm greeting allowed the Epsilon 1 population to publicly demonstrate their acceptance to a new way of life. In fact, the Epsilon 1 population was anxious to see the family's reaction when they saw where they were going to live and the technologies that they were about to experience.

Although Jason did not approve of having the press intrude on the privacy of the family, he realized they had become celebrities. And, unless the public saw them at this time, they wouldn't stop hounding him until they did. So he reluctantly agreed.

The plan was for the family to visit Jason's home module as a first step. In preparation for this, the press was allowed to set up cameras at hidden locations. The fact is, Jason, his wife Jennifer and Stewart were also anxious to see the family's reaction. Sort of like seeing the reaction of a child opening a Christmas gift.

It was planned that the arrival at Jason's home module would take place at mid-afternoon, and after describing what the family would see in the home module, they would spend the night. The next step was for Stewart to show Wanaka their home module the next day, with other family members going to their home modules at the same time. Jason and Stewart had done a lot of work helping to design the new modules so that the families would feel at home; such as having replicas of familiar furniture and other items made as part of the home module style. The landscaping was made to look like their former home planet, and the holographic imaging was recorded from surroundings that were familiar to them.

Because a fence had been set up to block the Epsilon people from seeing the Epsilon 1's common area while they were being rescued from their planet's flood, this was the first time for the family to see the wonders of a man made paradise. Although the people on board Epsilon 1 had become used to the common area, to a newcomer, the views would be breathtaking.

Using his translation device, Homanz asks Jason, "Why you have kept this from our people?"

"Homanz -- we made a decision at the beginning of the rescue to not expose your people to our technology. There is an old saying, that what you don't know can't hurt you or make you envious. If the Epsilon people knew of our technology they would probably want it for themselves, and it would make their lives less content."

"I understand, Jason, but this tells me you not know true nature my people."

"Homanz, you are probably right, but first we need to travel to my home, on board this train."

While on the Maglev train, the Epsilon family expresses amazement at the quiet speed and luxury they were experiencing. Upon arrival at Jason's home coordinates, they travel by way of an elevator to Jason's home level, and enter what appeared to be a wonderland. At one end of the 800-foot diameter by 100-foot-high living space was a magnificent house with beautiful landscaping. On each side of the house stood detached buildings. At the far end, opposite the house, a magnificent ultra-high definition view of the Pacific Ocean could be seen. Aside from the actual size of the living space, everything appeared to be infinitely expansive. The ocean view was one of looking from a high cliff, with tables, chairs, fire pit, barbeque, and kitchen appliances arranged for dining and relaxing activities. From this cliff side vantage point a person could smell salt air and listen to the sounds of crashing waves and seagulls. Between the house and cliff were perfectly manicured pathways, gardens, and trees, intrinsic to Earth's Pacific coast. At the center of the home module was a large uniquely designed swimming pool. The pool had a white sand bottom and two bridges passing over streams that connected two smaller sized pools.

As the Epsilon people looked in awe at the magnificent sight of Jason and Jennifer's home module, an android approached them, driving an open electric cart. The android asked six of the group to get in the cart for the trip to the house, and before leaving, said that he would come back for the remaining four on a second trip. Stewart, Wanaka, and Wanaka's mother and father stayed behind and advised the android that they would walk to the house.

The house was well appointed and functional. It was an upside down house, in that five bedrooms, with separate baths were located on the lower level, while the kitchen, dining room, and living area were located on the top floor to take advantage of the view.

In preparation for the visit, the android -- named Joe -- had prepared the five bedrooms to accommodate the family members, however, because the bedroom space was limited, the plan was for Jason and Jennifer to spend the night in the poolside cabana and recreation room.

After showing each family member their accommodations, Joe told them to meet by the ocean side cliff for dinner. He had prepared a dinner that included the planet's indigenous food recipes to help make the guests feel more at home. Before the dinner, everyone was offered wine and a chance for conversation. When they had all gathered, Jason stepped forward to speak to the group.

"I know each of you has questions, so this is a good time to do so."

Wanaka stands, "Will each our family have place like this?"

"Yes. Stewart, Jennifer, and I have spent much time designing each of your home modules, and I hope you like them. Homanz will live in his son and daughter-in-law's home module, but will have a separate house. Our plan is for each of you to visit your home module tomorrow."

Wanaka's mother says, "I see many things, but not know about them. Please explain how work."

"Yes -- let me begin -- they work using what we call electric power. Electric power, for all that you see, is provided by special boxes that receive fuel from pipes. This fuel is like coal or firewood, but we call it helium 3. In addition to the helium 3 fuel, we also receive air and water from pipes. Waste and trash is processed in each home module and sent in special containers to a central location using a second lifting device like the one you came here in -- we call it a service elevator. The android, named Joe, removes trash each day. Special containers of food and other needs, come into the module using this same service elevator. These needs are ordered by a device that senses that we are running out. For instance, if we need a container of salt, the salt holder senses that it needs to be replaced with a filled container. When the new container arrives, Joe places it into one of the salt container holders and removes the empty container and sends it back for reprocessing. The empty salt container is then cleaned and returned to the salt refill location. In other words, all food containers are reused so there is no resulting trash."

With a puzzled look on his face, Wanaka's father asks "how you move house and big things to -- what you call -- home module."

"We use a third elevator for big things. Our house was made of smaller parts and fitted together in here. We design all large things to fit into the large elevator. These parts are also designed for disassembly after the house becomes old. We use this method for almost all of the things that you see, so that everything is reused. When disassembled house parts cannot be reused to remake a new house, the materials are reworked for other purposes."

Wanaka's cousin points to the android. "How work Joe?"

"Joe is like a human, except he is mechanical, and operates using electricity. Joe can work 24 hours per day. During the day he does what you see him doing now, but at night he spends most of his time working on our landscape. When flowers die he replaces them. When leaves on trees die he clips them off. When weeds grow he plucks them out. When grass is too long he cuts it. When trees get too tall he uses a machine to cut branches. When trees die, he orders new trees and cuts the old tree for making wood products and mulch."

Wanaka steps forward. "How look like ocean?"

Jason smiles. "I assume you mean the view that we are now looking at. It is an artificial image, but as you can see it looks very real. Jennifer and I like this image very much, but we are planning to have a new one soon."

"What new image like?"

"Jennifer and I have decided to do something like one of our neighbors. It's an image that simulates the four seasons on our home planet that we call spring, summer, fall, and winter. It will be a mountain view from a place called New Hampshire.

Since we have almost unlimited fuel for electric power, our home modules can simulate temperatures and conditions without concern for running out. So in the four season image, we will have lower temperatures in the spring, winter, and fall, including rain in the spring, and snow in the winter. Because our current landscape design is for a mild climate, we will replace everything that you see, like palm trees, with maple trees and grassy meadows. Even though we can easily create the artificial image now, we need to spend time modifying the landscape to match the image."

Homanz wrinkles his brow. "What work people do?"

Jason nods. "Good question. Everyone who is able to work, is required to work two days each week, but can work more if they want to work more. Almost all manual labor is done by androids -- like Joe. So almost all human jobs are to supervise androids. Everyone receives the same pay credits whether they work more hours or have a highly skilled profession -- like a doctor. Since everyone can have almost anything that they want, a pay credit limit prevents abuse."

Homanz asks, "How you decide good worker from poor worker?"

"Homanz -- that's a very good question. We have twelve sectors--each having a large city -- or what you might call a village. Each sector has a yearly competition for awards. Awards can be for excellent products or service, and includes movie production, sports and school competition. The top sector gets a special recognition each year. This recognition includes having each family getting extra credits toward things like a more frequent remodeling of their home module."

Wanaka sets down her drink. "So what people do when not working?"

"Wanaka, we work to better ourselves. What I mean is -- we learn as much as we can and do creative things -- we also exercise and play sports to improve our health."

"What you do?"

"As you can see I have, what I call a creativity workshop -- located to the right side of our house. In this workshop, I have woodworking and metalworking tools. I also have an exercise room and a place to write books. To the left side of the house is another creativity workshop for Jennifer. Jennifer likes to grow things like flowers and vegetables. She also likes to compose music and paint."

Wanaka responds, "What Stewart do?"

"I can't speak for Stewart, but I know that he plans to spend one more year at the university to get his doctor's degree in science. After that he will probably work at the central science laboratory with me. And, if I know Stewart, he will become consumed in his work, and like me he will work more than just two days a week. What do you say, Stewart?"

Stewart pauses for a moment. "You're right about going back to the university, but I don't know about being consumed by my work. Right now, I want to spend as much time as I can with my family."

Jason looks around the room. "Are there any more questions before we have our meal?"

Wanaka's cousin raises his hand. "What kind jobs we have?"

"There are many types of jobs from which to choose. Since you are familiar with hunting, farming, and fishing, you might choose to supervise one of our wild animal preserves, cattle ranches, food growing facilities, or fisheries. Ok let's sit down to eat. We can talk more then."

MY COMMENTARY

If you want to know what Jason and his crew found on the Earth-like planet and the subsequent events that took place – look for my next book that will probably be titled *Epsilon Eridani*. The main point I'm making here is to describe how Earth's humans might accommodate a growing population by living in space and eventually traveling beyond our solar system in multi-generational spacecraft. To achieve traveling at a speed capable of reaching Epsilon Eridani in only one hundred ten years takes a lot more imagination because accelerating and decelerating to and from $1/10^{th}$ the speed of light (18,600 miles per second) would take an enormous amount of fuel. If we do eventually travel to the stars, we will probably need a new, yet to be discovered, propulsion system.

With regard to the Epsilon Eridani story, I thought of ants and bees that relentlessly pursue a single task. Isn't this what we could do with millions, or billions, of self-replicating and programmable robots working in space and building huge satellite/spacecraft? If I had the resources needed to develop a movie, I would visually show how this could happen. I would begin the movie with a zero-gravity space factory in which materials from asteroids are sorted into their various types. Included in the factory would be a sub-factory that continuously creates millions of specialized robots. Using the asteroid materials, the specialized robots would operate other sub-factories that, among other things, make tunnel-boring machines and cargo-carrying spacecraft. The movie would show millions of these spacecraft and boring machines relentlessly excavating large asteroids and returning materials back to the space factory, like a colony of ants, or swarm of bees going to and from their assigned objective. As part of my movie, I would show some of the cargo carrying spacecraft traveling to Jupiter to collect and freeze helium 3, while others travel to and from water-containing asteroids.

At the back of the space factory I would illustrate how a huge satellite/spacecraft could be constructed. As part of the subassembly process, I would show how we could construct the 800-foot diameter by 100-feet high living quarters. After welding a steel frame, with a honeycomb surface, residual asteroid material "regolith" and water would be used to make a form of concrete that could be pumped into the honeycomb voids. Although the 800-foot diameter by 100-foot high living quarters may seem huge by human standards, they would look like tiny specks while being used to construct the gigantic 16-mile diameter by 48-mile long satellite/spacecraft. By relentlessly assembling the living quarters, the desired object would eventually take shape.

Using today's special-effects cinematography, I believe that what I've just described could be shown in a way that would result in most viewers believing that what I am proposing is not so farfetched. The technology is not too difficult for most observers to understand and accept. As they say, a picture (in this case a movie) would be worth more than a thousand words.

As previously mentioned, one purpose in telling the fictional Epsilon Eridani story was to show that overpopulation of Planet Earth need not be a problem. With the potential for housing 30 trillion people in luxurious living conditions in space, I don't think we need to worry. However, we do need to worry about the short-sightedness of career politicians who can't see beyond the next election, and who don't take appropriate actions when needed. Another, and more profound, purpose for telling this story is to show how people in the future, through advanced language and communication capabilities, develop a commonly held religious belief.

With the entire Earth population living on satellite/ spacecraft, and having the ability to communicate and debate the big issues such as the origin of life, religious beliefs, and human purpose, a commonly believed truth is likely to prevail. This is not what we see happening today. With political correctness, media bias, corrupted entertainment, atheism, secular immorality, multiple religious beliefs, cultural differences, language problems, and a majority of Earth's people struggling just to survive, we currently have chaos. As a result, communication, debate, and logic generally fall on deaf ears, or in the case of Third World people, on no ears at all.

INTRODUCTION

Mark Twain once said, *"I admire those who seek the truth, but I am wary of those who claim to have found it."* Although I have yet to find a complete truth, through my life experiences and research I've discovered some things that may challenge what you may currently believe.

At one point I had considered making the title of this book *My Search for the Truth*, with a subtitle "What God May Have Had in Mind When Creating the Universe." I backed off from this title, worried most people would say that I was being too presumptuous. Who am I to know what God is thinking? On the other hand, I've delved into a lot of subjects related to who we are and what our purpose in life might be. As a result, I believe a lot needs to be said in spite of whatever controversy it might raise.

Most religiously oriented people believe we humans have been set apart from the animal world because we have souls, and our souls will leave our bodies when we experience a mortal death. Most of these people generally believe that when the soul leaves the body it will live on for eternity, with some going to heaven and others going to hell. How the soul lives this eternal life is vaguely described in the Bible. Since the Bible is not very explicit, I've often wondered about what eternity might be like and why it's our destiny. What is God's eternal plan, and why was the universe created in the first place? I'm especially interested in why God created humans who live in what some have called *"quiet desperation."* The great question is, why would God allow so many humans to suffer so much?

I don't just take things at face value. If there is a God or an advanced extraterrestrial race influencing human affairs, I want to see proof. In my opinion, it's unfortunate that most scientists don't want to implicate God in their research. To do so would say that what appear to be miracles are just things that we currently don't understand, but eventually will. Besides, who wants to be called a religious nutcase and have their government research funding cut?

In this regard, I'll quote famed physicist, author, Nobel Prize winner, and recognized genius Richard Feynman, who once said re: Feynman Quotes on atheists,

> *"God was invented to explain mystery. God is always invented to explain those things that we do not understand. Now, when you finally discover how something works, you get some laws which you are taking away from God; you don't need him anymore. But you need him for other mysteries. So therefore you leave him to create the universe because we haven't figured that out yet; you need him for understanding those things that you don't believe the laws will explain, or why you only live to a certain length of time – life and death – stuff like that. God is always associated with those things that you do not understand. Therefore, I don't think that the laws can be like God because they have figured it out."*

While I don't claim to be as smart as Richard Feynman, I can think for myself and draw my own conclusions. And so should you. I question everything, including the conclusions reached by brilliant scientists like Feynman. In my questioning, I began with the premise that there is a God until proven otherwise.

Too many things I've observed could not be a result of happenstance. And, the more questions I ask the more I'm convinced of this. Therefore, I examine everything in terms of there being a purpose behind all that there is. Who or what caused the *"Big Bang?"* Why is the Moon just the right size and location to facilitate life on Planet Earth? Is there a reason for the vast quantities of energy-producing helium 3 in the atmospheres of the gaseous planets? Why are solar systems so far apart? Why are asteroids gravitationally trapped between Jupiter and Mars so that they don't bombard us? And, are the asteroids there to serve a purpose? Why is the human population on Planet Earth just now starting to grow exponentially, but didn't grow exponentially for the past three thousand years? Why didn't some animal species, like buffalo, increase their population exponentially, if they survived the supposed extermination of dinosaurs 65 million years ago? How can an immensely complex living cell form by itself? How can living cells grow arms, legs, lungs, digestive systems, eyes, and ears? More importantly, how can living cells suddenly, or gradually, become male and female? How does the Earth's atmosphere maintain a 20 percent oxygen level for millions of years when just a small increase would cause the atmosphere to burn with the first lightning strike? What is gravity, and the electromagnetic and nuclear forces that hold matter together? How and why does time slow down with increasing speed? What is the so called "dark matter and energy" that's causing the expansion of the universe to accelerate, and is there a reason for their existence?

If you are a scientist who doesn't believe in God, these questions, and many others are, as Richard Feynman said, just unexplained mysteries that humans will eventually answer. To me, they indicate a creator with a purpose. Perhaps all will become known to me in an afterlife, but to believe that they can be explained as random acts of nature is the height of human arrogance. Feynman's argument that science will explain all of the aforementioned questions is quite frankly absurd unless God is part of the answer.

So what does this preamble have to do with this book? To me, as an engineer and quasi scientist, hypothesized answers to the above questions, and many more, are like pieces of a giant cosmic puzzle. When I try to assemble a puzzle, I begin by separating the border pieces and corners to establish a frame for the picture. Upon completion of the border, I separate the interior pieces into groups that appear to have the same object represented on them. I would try fitting these lesser number of pieces together until subcomponents of the picture start to emerge. At this point, if I see maple trees and a sky with clouds, I know that it's an outdoor scene in a temperate climate and summer season. If I see a familiar object like a house, I can begin to surmise what the overall picture might be and what it definitely is not.

It's my contention that God wants humans to see the picture on the cosmic puzzle box by piecing together the evidence that's been placed before them. Rather than ask why solar systems are so far apart, one might hypothesize what the reason could be. Perhaps the reason is we may not be ready at this point in our human existence to know what's there. When I step back and examine the United States history, it becomes evident to me that a series of near-miraculous events have brought America to where it is today. When I look at America's Constitution, economic system, topography, climate, diversity of people, form of government, and natural resources, I see purpose, not chance.

I don't see the Great Lakes as just an interesting and beautiful resource to look at and enjoy. Instead I see them as serving a purpose, like providing water to water-starved locations. When I see the Sonora and Mojave Deserts I don't just see their dramatic beauty, I also see how they could be transformed into greater beauty and help feed the world's growing population. When I see a vast open space between the populated coastal states, I see a place where great new cities can be created that will inspire the world. When I see people from all parts of the globe living together, I don't see cultural incompatibility; instead, I see a merging of cultures and backgrounds that will, in time, serve a common purpose. A common purpose that may eventually unite all of the world's people. When I see a form of governance that provides personal opportunity and freedom from tyranny and bondage, I also see an example for the rest of the world to follow. When I see economic growth and prosperity resulting from a free market system, I also see a means by which all of God's children can eventually prosper and fulfill God's plan for humanity.

Another one of my favorite quotes from Mark Twain is, *"It's easier to fool people than it is to convince them that they have been fooled."* P. T. Barnum supposedly said it more succinctly, *"There is sucker born every minute."* Don't become part of the lynch mob. Question everything, and when you find what sounds like a logical answer, ask the next question until you have a good in-depth understanding about what you believe to be true. I've found through my investigations that mainstream thinking is almost always wrong, especially about subjects that can't be proven scientifically. Just look at the issue of manmade global warming. Don't be blinded by cognitive dissonance. Don't believe simply because you think the persons espousing a scientific premise are smarter than you. I've known lots of smart people who earned their credentials by having a good memory, but had difficulty thinking logically. Many of them were apparently so locked into memorizing a subject, they lacked an ability to thoughtfully question what they'd memorized.

Because I'm an engineer, I tend to have a sense of logic, as well as a strong degree of skepticism about almost everything. Often, what first seems logical, later turns out to be illogical, when further probing the subject. For many years of my life I believed what I was told about Darwin's theory of evolution. The advocates showed how various species adapted to environmental changes (a true form of evolution), but this aspect of evolution tricked me into believing that species themselves can also change over long periods of time. The fact is, the evidence for this claim has never been found. In my opinion, this evidence will never be found; and if it is, it will turn out to be a hoax.

Since I'm currently in my seventies, I've had a long lifetime of experiences, and in retirement I've had a lot of time to reflect and ask questions, especially about God and how engineering ideas can economically help the poor and destitute people around the world. One thing I've found is how new and original ideas are almost always met with some opposition. With this introduction and the prologue, you already know a lot about my current and evolving worldview. As stated inside this book's cover, I caution you about reading it because you may experience cognitive dissonance. However, if you do continue reading I guarantee you will be surprised by what my research has found.

In conclusion, I'd like for you to keep in mind the words of one of the leading figures in quantum physics, Niels Bohr, *"The problem with your idea is not that it is crazy, but that it's not crazy enough."* Bohr's point was that reality has shown itself to be stranger and more bizarre than science fiction.

CHAPTER 1

WAS THERE A BIG BANG?

'm beginning the first chapter of this book with a discussion about how the universe and life began, because it illustrates how mainstream science is desperately trying avoid implicating God in all that we see around us. This avoidance of the obvious is both amusing and disturbing to me. It's amusing because science will reluctantly discover God as time passes, but it's also disturbing because so many people are being hoodwinked into a false belief. However, in their hopeless pursuit, science is allowing us to examine an alternative to God in a way that indirectly confirms what many of us have come to believe as truth. It takes away the brainwashing and lack of scientific thinking that might occur if only the theologians were in charge. Knowing that the mainstream scientific community has an anti-God bias, we can rest assured that no stone will be left unturned in their relentless attempt to undermine God's work. And, it's this relentless pursuit that will eventually reveal God's intent and purpose.

HAWKING'S THEORY

I recently watched a Science Channel telecast entitled *An Introduction to the Universe*. The program is presented in the words of Stephen Hawking and subtitled as the story of everything. As most of you may know, Stephen Hawking is regarded by many in the scientific community to be the Einstein of our time.

The telecast began with an explanation for how the universe came into existence. At first there was a very small atom-sized entity called a singularity, and time and space as we know it did not exist. Then the singularity exploded, and within a trillionth of a second it was the size of an orange. In one hundred seconds it expanded to a diameter that was billions of miles across. In ten minutes it was thousands of light years in diameter. Contained in the subatomic radiated particles from the explosion were matter and anti-matter, which annihilated each other when they came into contact. Luckily for us, according to Hawking, an imbalance of one in one billion particles was matter that survived the blast.

Three hundred and thirty-thousand years later, a subatomic particle fog lifted and became visible as hydrogen atoms. At this point, gravitational attraction and naturally occurring spatial imperfections took over and caused the hydrogen particles to cluster together. As this clustering took place, a fusion reaction occurred as the impacting forces of the hydrogen atoms reached a temperature of ten million degrees, thus forming stars like our sun. At the center of these stars, the hydrogen atoms fused to form the element helium; but because of tremendous gravitational forces, the temperature further increased to cause additional fusion reactions of the helium atoms, which resulted in the formation of carbon.

Finally, the carbon atoms fused to form iron atoms. Because the iron atoms could not continue the fusion process, the stars' fuel supply ran out and they began to collapse in on themselves and cause what's called a super nova. But, because of the collapsing process, higher temperatures were reached and the iron atoms began to fuse together. As a result, all of the heavier elements that currently exist were created and new stars and solar systems began to form from the exploded debris. These solar systems were also affected by gravitational attraction and their clustering resulted in galaxies. At the center of each galaxy is an unrestrained gravitational collapse that resulted in what is called a gigantic black hole, where the gravitational attraction is so great that even light cannot escape.

Within one of the billions of galaxies that were formed, we find an instance where a planet called Earth had the mysterious special characteristics needed to form organic carbon based life. Some of that life mysteriously developed into a special type that had the ability to contemplate what caused the universe and everything in it.

If the above description of the formation of the universe and human life isn't mind boggling enough, the wonders and mysteries of the universe continue even further. Scientists now say that the universe is not only expanding, but that the expansion is accelerating. This scientific finding runs counter to intuition, which would predict a slowing down and a reversal of the universe as a result of gravitational attraction. What this finding implies is that a mysterious substance called "dark energy" is causing the acceleration. If that were not strange enough, other observations of galactic motion infer that another substance called "dark matter" must also exist. When working out the cosmic math, scientists have concluded that 70 percent of the universe is dark energy and 25 percent is dark matter, thus only 5 percent of the universe is matter which is visible to our current detection devices.

Because we can see, with our telescopes, exploding stars in our own Milky Way galaxy, we can infer that stars and their solar systems have a finite life. If stars have a finite life, and their exploding debris is insufficient to form new stars, it's likely that the universe could eventually burn up and become space, littered with black holes. If the dark energy that scientists claim is driving the universe apart runs out of steam, the universe could reverse its expansion and collapse in upon itself due to gravity – a process that Hawking calls *"the Big Crunch."* If this were to happen, the universe could return to something like the Hawking singularity and explode again in a never ending cycle of universe recreation. In other words, our universe will probably end. And, with the universe ending, so will what we might call eternity -- end as well.

With these elaborate phenomena in mind, it makes me ask, where does God fit into all of this? In his narrative, Hawking poses the question that he is often asked, is there a designer? His answer is *"not necessarily so,"* and his reasoning is that there may have been an infinite number of other universes (multiverses), and only one may have allowed Earth's coincidences to happen. Apparently Hawking understood the mathematical impossibilities for all these miracles to work out just right, and an infinite number of universes argument tends to counteract these impossibilities!

One other point worth mentioning is Hawking's claim that an asteroid is likely to hit planet Earth sometime in the future and wipe out most of Earth's life forms. He points to the asteroid that hit the Yucatan Peninsula sixty-five million of years ago and supposedly wiped out the dinosaurs as evidence that it will probably happen again. Because of this possibility, Hawking warns that humans need to develop a capability for living on Mars and eventually to interstellar migration.

WAS THERE A BIG BANG?

It's interesting to note that during Hawking's Science Channel program, an interstellar spacecraft was graphically illustrated. The spacecraft was shown as a group of cylinders arranged in a circle, with one larger cylinder at the center that acted as the propulsion system. It wasn't clear how large the spacecraft was or whether there was rotation to simulate gravity. In any event, it was an attempt to show how human generations could live in space while traveling to other solar systems.

As an aside, it is somewhat curious to note that Hawking claims that in ten minutes after the Big Bang explosion, matter and antimatter were thousands of light years from the originating singularity. Does this mean that matter had traveled faster than the speed of light, or does it mean that the speed of light was faster at this early point in time? Hmm . . . interesting! As you know, the speed of light is considered to be a constant and much of what we assume for determining age is dependent on this assumption.

So there you have it, *"the story of everything,"* as told by cosmology's most prominent spokesperson. But is there another explanation that gives meaning and purpose to it all? While I am not scientifically qualified to render a story of everything, I will give it my best shot. However, my story of everything will be based upon having a God that has designed and implemented all that there is, and that it was done for a purpose. Since I may not be as smart or as scientifically trained as Hawking, you might regard the following as a minority opinion; after all, a few scientific facts do not constitute a whole. There is plenty of room for interpretation.

My story of everything begins with vast amounts of matter (let's call it "dark matter") and the existence of God. The matter then converts to energy (which we can call "dark energy"). When all of the matter is converted to energy, it converts back to matter and the process begins again.

In the process of converting matter to energy, God converts some of the dark matter into elaborately designed particles. These created particles, only five percent, form the visible universe. In other words, these particles are the galaxies and solar systems we are currently able to see and measure. For want of a better explanation, these galaxies and solar systems are formed in a manner similar to what Stephen Hawking has claimed to have happened. My guess, however, is that God's created particles (atoms) have command and control capabilities, whereby God can direct their behavior, like creating Adam from the dust of the ground.

My contention is that God allowed the galaxies and solar systems to form in a random and aesthetically beautiful fashion, much like the beauty of a fireworks display. But, the real beauty and creativity of the universe probably comes from the planets contained within the solar systems. Whether or not the beginning or ending of the galaxy and solar system forming process requires billions of years is not important. What is important is that there be a window of time where biological life can exist during the solar system's and universe's existence.

Regarding the end of the universe, as mentioned above, the galaxies and solar systems will eventually become black holes. And, while I haven't yet mentioned the fate of black holes, I'll describe their fate here. Black holes eventually vanish as they emit gamma rays into space. The gamma radiation is called "Hawking radiation" in recognition of his discovery. Since the gamma rays are a particle form, I'll assume that they are designed to eventually become dark energy to complete my proposed dark matter to dark energy cycle.

Hawking's *"story of everything"* model is based upon a universe that just happened by accident, while my "story of everything" model is based upon a universe that was created for a purpose. Hawking's model starts with a singularity that is no bigger than an atom, while my model assumes conservation of mass and energy. Hawking further assumes that matter and antimatter came into existence first, but an imbalance favored matter and kept the universe from completely annihilating itself. While I have no equivalent matter and antimatter annihilation process, in my model, I will acknowledge that antimatter can scientifically exist, but it can be manufactured by humans for a purpose.

Hawking doesn't say how or what started it all. When secularists are confronted with this question, they might counter by asking how did God come into existence? From my point of view, it's easier to answer the secularist question. Since God is a spiritual being, I don't have to explain how eyes, ears, hearts, stomachs, other mutually dependent body parts and the reproductive process came into existence through natural processes.

I find the existence of dark energy and dark matter to be quite intriguing. Can the dark energy be harnessed to make distant space travel feasible for mortal—or immortal—humans? Can dark matter be used as a propulsion fuel for space travel? Does the existence of dark energy mean all things, including thoughts, in the universe are interconnected? Does dark matter or energy serve as a communication medium for God and other godlike beings (immortals or angels)? By asking these questions, we can at least begin to make some sense of it all.

Let me stretch your imagination some more. If we can eventually create huge rotating spacecraft, as described in the Prologue, using millions of self-replicating robots, is it possible to robotically create a Moon-sized spacecraft? What's the limit? I pose this question because, interestingly, the planet Venus has a density and size similar to Planet Earth. By using a spacecraft Moon as a gravitational "tugboat," Venus could be maneuvered into Earth's orbit and have its spin rate adjusted to a 24-hour day. I'll leave you with this thought and let you decide what it may mean. Besides, who knows what humans will be capable of doing hundreds or even thousands of years from now. How about creating an Earth-like planet in six days!

DARWIN'S THEORY

I remember years ago when I went to church only on Easter and Christmas and believed in Darwin's theory, I asked my church-going mother-in-law if she believed in Adam and Eve. When she stated that she did, I was dumbfounded. How could anyone believe in something so absurd? As time went by, I was confronted by two missionaries who were passing out religious pamphlets. Not to be unkind by rejecting the pamphlets, I accepted them and bid the missionaries farewell.

Although I wasn't interested in reading the pamphlets, one title did strike my interest. It had to do with why Darwin's theory was wrong. As I began reading, it occurred to me that what the pamphlet was saying seemed logical. For some reason I never thought about why cats can't mate with dogs and vice-versa. If this were true, then how did one animal species evolve into another species, as I was taught to believe? Without going into detail about other aspects of the pamphlet, it had a profound effect on my thinking at that time. Could it be that what I had taken for granted was provably wrong?

At this point, I decided to read a book that supported evolution. I can't remember the book's title, but while reading the author's description of how the first living cell was formed, I was flabbergasted. How much illogical nonsense could I take? However, I didn't give it much thought until years later when I read Dr. Walt Brown's book *In the Beginning*. His logic and scientific analysis were profound and everything about the Bible's description of creation and the flood began to make sense to me. And, it was at this point in time that I remembered my mother-in-law's belief in Adam and Eve. Perhaps she was right after all. And I told her so! More about Dr. Brown's book is presented in my Flood chapter commentary.

So why do so many people still believe in Darwinian "macro" evolution (not "micro" evolution, or adaption, which is a proven scientific fact), when a little study and common sense would show that it's illogical nonsense? Well, most people haven't studied it, and those who espouse it as fact find it a convenient excuse for saying there is no God. With this opening, I'll give two very important examples of why Darwin's theory is a house of cards. The first example deals with the fossil record and the second deals with DNA.

Example number one. In 1909, Dr. Charles Walcott found some shale rocks that yielded the most important fossils ever found. These fossils revealed that complex life suddenly appeared supposedly about 530 million years ago, the Cambrian fossils.[1] Contained in these fossils were all of the existing and extinct species ever known, without any evidence of there being transitional species. As a result of this finding, the evolutionist's theory of change in species and gradualism was discredited. However, at that time Darwin was being raised to near sainthood, and no reputable scientist questioned the roll of random gradual evolution. Walcott was the director of the Smithsonian Institution, and a world-renowned paleontologist. Although he collected more than sixty thousand fossils, he reburied them in the basement drawers of his laboratory. There they stayed for eighty years before they were rediscovered and made available to the scientific community and to the public. Even so, authors of high school textbooks and introductory courses in college biology still ignore these data. Why? The reason is that the data runs counter to the logic of gradualism and the paradigm they learned in their youth.

Example number two. DNA. With regard to creating a self-replicating living cell by random acts of nature, I offer the following as only one of many explanations of why life could

---

1 *The sudden appearance of fully formed fossil species (eyes, lungs, digestive systems, etc.) These fossils represent every species that has ever existed. Since no transitional fossils have been found, scientists have difficulty explaining how they could have formed so quickly from single-celled organisms. Also, since radiometric dating indicates that all currently known species formed during the Cambrian time period,* supposedly *about 530 million years ago, this finding tends to cast doubt upon the macro evolution theory that requires billions of years for fully formed species to develop.*

not have evolved on its own, without the omnipotent power of God. The double helix DNA molecule uses a four character "alphabet" to create 64 three-letter "words," including "start" and "stop" to separate "sentences." Depending on how the letters and sentences are strung together, this one molecule creates all living things. Before it can create a living creature, it must communicate this language to proteins that are made up of twenty amino acids. Proteins have their own twenty-letter alphabet; one letter for each amino acid. These letters combine to create hundreds of words. The problem is that the languages are mutually exclusive. What's necessary is a translator called enzymes. Nobel Prize-winning physicists Francis Crick and James Watson claim that a protein is like a two hundred-letter paragraph lined up in the correct order. And, the possibility of this happening by chance is a number so great that it is beyond imagination; comparable to all the atoms in the universe.

If you want to learn more about what I've just said, I recommend that you read *Signature in the Cell – DNA and the Evidence for Intelligent Design*, By Stephen Meyer PhD, and *Undeniable – How Biology Confirms Our Intuition that Life is Designed*, by Douglas Axe PhD. Folks – it doesn't get any better than this when trying to prove God's handiwork. DNA is unquestionably acting like a biological computer program that directs the formation of living cells and eventually the resulting features and traits of all animals or humans. This fact alone justifies my using the words *"undeniable proof"* of intelligent design and the fact that science is discovering God.

Because of Crick's and Watson's findings, most of today's informed scientific community has generally given up on trying to prove Darwin's macro evolution theory. In a tenacious attempt to avoid giving credit to God, alternate theories such as panspermia and multi-universes are now being espoused. It's my opinion that the Darwinian theory of evolution will itself evolve, or change, as time goes by. It will then be superseded by the theory of extraterrestrial intervention and perhaps one or more additional theories. But, in the long run, I believe the final theory will have no recourse but to nominate God as the creator of all things.

SCHROEDER'S HYPOTHESIS

If you do believe in God (or even extraterrestrials) you might conclude that the Cambrian fossils were created by God more than 500 million years ago, and allowed to develop into what we see today. In this instance, you have a case that has some credibility, and that's what Dr. Gerald Schroeder says in his book *The Science of God*.

Schroeder makes the case for currently known science and the Bibles book of Genesis merging to say the same thing. He claims that the first five days of creation were not twenty-four-hour days as we know them, but rather they were eras that coincide with the Big Bang theory. One basis for his hypothesis is Einstein's discovery of relativity and time. If one were to travel near the speed of light, time would slow down to near zero as billions of Earth years pass. Thus, during the supposed fifteen-billion-year-old universe, about 530 million years ago, on the fifth day God created fish, birds, and animal life; the Cambrian fossils. Of course, on the sixth day God created Adam and Eve. From this point forward, Schroeder claims that days are twenty-four hours as we know them. Schroeder explains this divergence by referencing the Bible's original Hebrew text to explain the misinterpretation of the word *day*. He goes on to explain such things as before the first day of creation there was no such thing as a day, and time and space didn't exist.

Schroeder's interpretations of Genesis' six days of creation falls in line with current science, including the Cosmic Microwave Background (CMB) discovered in 1964 by Nobel Prize winning physicists Arno Penzias and Robert Wilson. Their finding was based upon a residual afterglow in dark space that can be detected in the microwave portion of the spectrum, and concludes that the universe began with a big bang which happened fifteen billion years ago. Schroeder states that before 1964 scientists believed that the universe just existed and there was no beginning.

My problem with Schroeder's hypothesis is that current mainstream science is changing all the time, and to take a snapshot based upon today's scientific knowledge can be misleading. However, with that being said, Schroeder does give credit to God for designing and creating all that there is.

CHAPTER 2

ARE THERE UFOs?

Famed Manhattan Project scientist Enrico Fermi asked the question, *"Where are they?"* He asked this question when he and his colleagues were having a side conversation while developing the atomic bomb. Fermi noted that if there are billions of planets in the universe with extraterrestrials that are billions of years more technically advanced than us living on them, they should have contacted us by now. Perhaps, based upon this question, we should ask, *"Are we alone?"* I believe that the existence of Planet Earth is in itself a miracle. Too many fortuitous things indicate that a form of intelligence has made Earth possible and it appears to me that we are alone in the universe, with the exception of God and His helpers, which at this point I'll call *angels*.

So in my search to help support or refute this premise, I'd like to again quote Richard Feynman re: Feynman quotes on atheists,

> *"It doesn't seem to me that this fantastically marvelous universe, this tremendous range of time and space and different kinds of animals, and all the different planets, and all the atoms with all their motions, and so on, all this complicated thing can merely be a stage so that God*

*can watch human beings struggle for good and evil —
which is the view that religion has. The stage is too big for
the drama."*

This quote is quite interesting and provocative, but it makes
one huge assumption. That assumption is that the universe
contains billions of Earth-like planets and God can't possibly
keep track of them all. It also appears to make the assumption
that God doesn't have helpers, but how about angels? Do you
believe that there are billions, or even trillions, of Earth-like
planets? After all, Earth is just a tiny speck in the vastness of
the universe, and logic would say that there must be more than
one Earth-like planet.

In my search for the truth and reinforcement of my belief in
the existence of God, I've given a lot of thought to the apparent
miracles surrounding our tiny Planet Earth. Mainstream
scientists currently believe that Earth is just one of millions of
planets in the universe that can support life. In fact, Dr. Frank
Drake, from Search for Extra Terrestrial Intelligence (SETI),
formulated the Drake equation that predicts the number of
these planets. The equation encompassed many variables such
as the number of stars that resemble our sun and the percentage
of these stars that may have Earth-like planets in habitable
zones. The problem with this equation is that it doesn't account
for recent scientific discoveries about the uniqueness of Planet
Earth. So, regardless of how many trillions of planets that exist
in the universe, I'll take the contrarian view that there are no
other Earth-like planets with intelligent life, and if there are,
it's because God created them. Here's my reasoning for this
point of view, which I'll bet will never be presented on the
Science or History Channels.

ARE WE ALONE?

In his fascinating book *The Case for a Creator*, *Chicago Tribune* columnist Lee Strobel interviews world-renowned experts in the fields of cosmology, physics, astronomy, and biochemistry, to determine whether current science is pointing toward or away from God. As you might suspect, the conclusion was that science is overwhelmingly pointing to a creator God. Strobel's evidence shows scientific information that has emerged over the last fifty years is revealing more and more unexplainable complexity, and the precision by which this complexity is formed can only be attributed to a master designer. The possibility for it to have been formed by random acts of nature is so remote that, to quote a passage from Strobel's book, *"it would be like throwing a dart from space and hitting a target on Earth that was a trillionth of a trillionth of an inch in diameter."*

In an effort to not plagiarize the wording in Strobel's book, I'll try to summarize what he has presented in my own words, but since his book states the case so well I am compelled use quotes where I cannot do better. To begin I'll use this quote:

> *"Earth's location, its size, its composition, its structure, its atmosphere, its temperature, its internal dynamics, and its many intricate cycles that are essential for life – the carbon cycle, the oxygen cycle, the nitrogen cycle, the phosphorous cycle, the sulfur cycle, the calcium cycle, the sodium cycle and so on – testify to the degree to which our planet is exquisitely and precariously balanced."*

Let's begin with Earth's size and location. In a recent television program entitled *The Universe*, a spokesperson stated (and I'm paraphrasing) that in the past, people thought the

Earth was the center of the universe; however, we now know that the Earth orbits around the sun rather than the sun orbiting the Earth, and that the sun is just one of billions of suns in the universe. The spokesperson went on to say the Earth is just an ordinary planet remotely located in an ordinary galaxy. Although this spokesperson was a PhD cosmologist, it seems odd that she wouldn't have been more aware of recent scientific findings. If she did know, her statements were deliberately false and misleading.

Is the Earth located in an ordinary location in an ordinary galaxy? The answer is no. In Lee Strobel's taped interview with famed astronomer Guillermo Gonzalez – informally known as the "star guy," here is what he said:

> *"Galaxies have varying degrees of star formation, where interstellar gases coalesce to form stars, star clusters, and massive stars that blow up as supernovae. Places with active star formations are very dangerous, because that's where supernovae explosions occur at a fairly high rate. In our galaxy, those dangerous places are primarily in the spiral arms, where there are hazardous giant molecular clouds. Fortunately, we happen to be situated safely between the Sagittarius and Perseus spiral arms."*

Also, we are very far from the nucleus of the galaxy, which is also a dangerous place. We now know that there's a massive black hole at the center of our galaxy. In fact, the Hubble space telescope has found that every large nearby galaxy has a giant black hole at its nucleus. And believe me, these are very dangerous things!

Most black holes, at any given time, are inactive. But whenever anything gets near or falls into one, it gets torn up by the strong tidal forces. Lots of energy is released, such as gamma rays, X-rays, particle radiation, and anything in the inner region of the galaxy would be subject to high radiation levels. That's very dangerous for life forms. The center of the galaxy is also very dangerous because there are more supernovae exploding in that region.

One more thing. The composition of a spiral galaxy changes as you go out from the center. The abundance of heavy elements is greater towards the center, because that's where star formation has been more vigorous over the history of the galaxy. So it has been able to cook the hydrogen and helium into heavy elements more quickly, whereas in the outer disk of the galaxy, star formation has been going on more slowly over the years and so the abundance of heavy elements isn't quite as high. Consequently, the outer regions of the disk are less likely to have Earth-type planets.

Gonzales puts it this way, *"Now put all this together. The inner region of the galaxy is much more dangerous from radiation and other threats; the outer part of the galaxy isn't going to be able to form Earth-like planets because the heavy elements are not abundant enough; and I haven't even mentioned how the thin disk of our galaxy helps our sun stay in its desirable circular orbit. A very eccentric orbit could cause it to cross spiral arms and visit the dangerous inner regions of the galaxy, but by being circular it remains in the safe zone; this all works together to create a narrow safe zone where life sustaining planets are possible."*

Because of numerous pages of descriptive information in Strobel's book, I'll try to summarize Gonzales' discussion about the unique nature of our Milky Way Galaxy. First of all, most galaxies are elliptical rather than spiral. As a result, the contained solar systems orbit their central black hole like a swarm of bees. Because of the wide variation of their travel, they are not conducive to supporting life. The second most abundant type of galaxy is called *irregulars*, and their solar system randomness is also not conducive to supporting life. Spiral galaxies are the least abundant types, but for various reasons, their variation in size also makes most of them unlikely to support life. This leaves only about two percent of the total number of galaxies with the capability of having an Earth-like planet contained within them. In the words of Dr. Gonzales, *"I've studied other regions—spiral arms, globular clusters, edge of disks—and no matter where it is, it's worse for life. I can't think of any better place than where we are."*

Regarding recent findings of planets circling other stars, Gonzales pointed out that most of the orbits were highly elliptical, with only a few being circular. This finding surprised astronomers, because they believed other planetary systems would be like ours. The problem with elliptical orbits is that they pose a problem for life. A planet with the mass of Planet Earth would be sensitive to the gas giant planets if they had more eccentric orbits. This eccentricity would result in making the Earth's orbit more eccentric and causing dangerous surface temperature variations.

If all of the above wasn't enough to place Earth in a special place, consider how perfectly situated Earth is relative to the sun to maintain a watery surface and how Jupiter acts as a space vacuum cleaner to protect Earth from being bombarded by asteroids and comets. And, as Dr. Gonzales points out, our Moon stabilizes the Earth's tilt axis. If our Moon were not there, our tilt axis would swing wildly over a large range and result in major temperature swings. For example, if our tilt axis were close to 90 degrees, rather than its current 23.5 degrees, the North Pole would be exposed to the sun for six months, while the South Pole would be in darkness, then vice versa. Fortunately, the Moon stabilizes the Earth's axis to a variance of about one and one half degrees. In addition, if the Moon were larger than it is, ocean tides would be too strong and global circulatory currents would not occur.

Then what about our sun; is there anything unique about it relative to other suns in the universe? Again the answer is yes. First of all, it represents ten percent of the most massive stars in the galaxy; which means if you were to pick a star at random, it would most likely be a red dwarf. To make a long story short, red dwarf stars emit most of their radiation in the red part of the spectrum; thus making photosynthesis less efficient. But, an even greater problem exists with luminosity. So in order to maintain liquid water on an orbiting planet's surface it would need to be much closer to the red dwarf star. The problem is that tidal forces between the star and the planet become stronger, to a point where the planet ends up in what's called a tidally locked state. This means that the planet presents the same face towards the star; thus causing unacceptable temperature differences between the lit and unlit sides.

But that's not all. Since the intensity of red dwarf solar flares is about the same as our sun, the particle radiation would be too intense to be conducive to life; one problem being a stripping away of an ozone layer. In addition, the lack of ultraviolet radiation would be detrimental to establishing oxygen in a planet's atmosphere. Conversely in the case of more massive stars than our sun, the problem of having excessively high levels of ultraviolet radiation would be detrimental to the formation of life.

With stars more massive than our sun, the problem, in addition to the ultraviolet problem, is they don't live as long. Stars a little more massive than our sun live only a few billion years whereas our sun is expected to last about ten billion years. As a result, everything in their life cycle happens much faster, like changes in luminosity. Fortunately, our sun is stable in that its light output varies by one tenth of one percent over its eleven-year sun spot cycle.

Besides the stability created by its near circular orbit, the sun has a high abundance of heavy elements compared to other stars of its age and location within the galaxy. According to Dr. Gonzales, the sun's metallicity may be near the golden mean for building Earth-like habitable terrestrial planets.

In the words of Lee Strobel, *"It would take a star with the highly unusual properties of our sun—the right mass, the right light, the right composition, the right distance, the right orbit, the right galaxy, and the right location—to nurture living organisms on a circling orbit planet. That makes our sun, and our planet, very rare indeed."*

Now, what about the size and mass of Earth; is there anything unusual or unique that should cause us to consider whether or not we are alone in the universe? Again the answer is yes. First, a terrestrial planet needs a minimum mass to retain an atmosphere. In Dr. Gonzales words:

"You need an atmosphere for the free exchange of chemicals of life and to protect the inhabitants from cosmic radiation. And you need an oxygen rich atmosphere to support big-brained creatures like humans. Earth's atmosphere is 20 percent oxygen – that's just right it turns out. And the planet has to be a minimum size to keep the heat from

its interior from being lost too quickly. It's the heat from its radioactive decaying interior that drives the critically important mantle convection inside the Earth which results in our protective magnetic field. If Earth were smaller, like Mars, it would cool down too quickly; in fact, Mars did cool down and is basically dead."

More massive planets, according to Dr. Gonzales, would create too much gravitational pull. This would result in a tendency to create a smooth sphere. If Earth were a smooth sphere, our oceans would cover the entire Earth to a depth of about two kilometers; thus creating a water world. In a water world, tides and weathering would not wash nutrients from continents into the oceans, where they feed organisms. Many of the life-essential minerals would sink to the bottom and the salt concentrations would be prohibitively high. Life can only tolerate a certain level of saltiness. On Earth, salt concentrations are maintained because of marshy areas along some coasts, which through evaporation leaves salt behind. In a water world, excess salt would settle to the bottom and would be inhospitable for life.

Besides having a convecting metallic liquid mantle below the Earth's surface causing a protective magnetic shield, it could be instrumental in causing heavy elements and ores, essential to the development of human life, to rise close enough to the surface to allow mining of essential elements.

While I've only touched upon the details supporting the serendipitous, finely tuned requirements and phenomena that established the Earth as a very unique place in the universe, I would like to mention one more very unusual fact: Earth's apparent setup for discovery. Let me begin by asking, don't you think that it's strange that when there is a solar eclipse that the Moon is the exact same size as the sun when viewed from Earth? It's even more unusual when you consider that, of the nine planets, with sixty-three moons, only Earth's moon provides a total eclipse. Moreover, by providing a total eclipse, Earth's humans have been able to learn about the nature of the stars using spectroscopes to confirm Einstein's theory of relativity by showing that gravity bends light, and allows astronomers to calculate the change in the Earth's rotation over the past several thousand years; which is important because it enables us to superimpose ancient calendars on our modern calendar system.

Dr. Gonzales points out that the Earth is finely tuned for humans to make measurements that are critical to understanding our universe. His examples include how the Earth is the best overall platform for astronomers and cosmologists to make a diverse range of discoveries. Our location away from the galaxy center and in a flat plane disk provides a privileged vantage point for observing both nearby and distant stars. Because of Earth's tilt axis, we maintain deep snow and ice deposits, which preserve historical data about the Earth's climate, its early atmosphere, and changes in the Earth's magnetic field. And then there is our atmosphere, which not only provides enough oxygen to sustain life, it also facilitates fire and the development of technology. Furthermore, it just so happens our atmosphere provides transparency, which would not be the case if it contained significant amounts of carbon-containing atoms like methane. A transparent atmosphere has allowed astronomy and cosmology to flourish.

Another interesting fact is that movement of the Earth's crust results in earthquakes and other phenomena critical to life. Using seismograph data, scientists have been able to provide a three-dimensional map of the structure of Earth's interior. Although not part of the discovery argument, without movement of the Earth's crust, we may not have continents and mountains that prevent us from having the previously described water world.

So there you have it, Earth needs to be precisely the right size, mass, and composition. The Moon needs to be precisely the right size and mass, and the sun needs to be precisely the right size, mass, and composition. In addition, they all need to be precisely located relative to each other, travel in near-circular orbits, and have a large planet nearby to shield it from asteroid bombardment. If that weren't enough, our solar system needs to be located in a specific type of rare spiral galaxy, and at a specific location within that galaxy. With that being said, the conditions on the Earth itself need to be just right to support life, and allow for human technological development.

If you were to visit another solar system that meets the same criteria, what do you think the odds would be that you would find a planet, moon, and planet shield that is precisely arranged according to Earth's specifications? Unless they were placed there deliberately, I think that you would agree that the odds against finding this rare combination are incredibly high. Needless to say, even if this rare combination were found, it doesn't mean that human-type life will have evolved on the Earth-like planet by random natural processes.

Given the billions of possible sites in the universe, do all of the above precise Earth-like requirements lead to the conclusion that Earth is the only planet of its type in the universe? If the answer is based upon random occurrences, I believe the odds favor Earth being a unique place, especially when you consider recent findings of orbital eccentricity. On the other hand, the unique occurrences themselves strongly indicate that intelligence is behind it all; and if so, that same intelligence could have arranged for many Earth-like planets. Needless to say, the phenomena that supports life and allows for technological advancement of the human species, is indeed miraculous in its own right.

Before ending this part of my commentary I'd like to briefly discuss NASA's current effort to explore the possibility of terraforming Mars. While terraforming Mars may sound like a commendable thing to do, its likelihood is off the map. Robotically manufacturing rotating spacecraft/satellites, or hollowing out existing asteroids makes much more sense. Mars is a dead planet that is traveling in an eccentric orbit far away from the sun. Without going into a lot of detail regarding the enormous obstacles involved in terraforming Mars, I recommend you check it out on the internet. From what has been discussed in this commentary, Mars, along with Venus and Mercury, are tangible examples of why the claims regarding Earth's unique location, size, and other attributes are true. They bear witness to the fact that, without specific direction, planets of random sizes and mass that are randomly distributed within the huge three-dimensional space of a specific solar system, will not result in the required Earth, moon, and sun orbits, and protective planet (Jupiter) combo. Change any one of these requirements, and the system fails.

While I haven't done the math, I believe that the statistical odds against this happening must overwhelmingly exceed the possible locations in the universe that this could take place. Besides the improbability of this combo occurring, the exact formation of an Earth-like planet's attributes is an even more compelling argument for uniqueness. Why, through random unguided processes, should we expect to discover the same type of atmosphere, with its stable 20 percent oxygen level and ozone protection? Why, through unguided processes, should we expect to locate a planet with a protective magnetic shield? Why, through unguided processes, should we expect to find a tilt axis with the same spin rate, that is conducive to stable temperatures, or near-surface ores and elements that have shaped human society, or organic life?

To add credence to what I have just said about the unique formation process requirement that results in an Earth-like planet, I checked out planet formation on the internet and found that science still knows little about planet formation. If all planets formed in the same or similar way, it could be argued that all planets would be similar to Earth; however, this is not the case. If we examine Mercury, we find it is denser, for its size, than Planet Earth, and much denser than our moon. While Venus is close to Earth's size and density, it has a very slow spin rate (243 Earth days per Venus day). It also has a carbon dioxide atmosphere that is equal to 90 Earth atmospheres; about the same as that found about one kilometer below our ocean surface. Also, if Earth's spin rate were the same as that of Venus, it would spend a lot of time facing away from the sun; thus causing extreme temperature differences that would not be conducive to human habitation. Although some scientists claim Venus' carbon dioxide atmosphere is the result of a runaway greenhouse effect, it's not clear why Earth's presumed early carbon dioxide atmosphere did not end up the same way (an oxygen free condition needed for single-cell bacteria to form). Instead, Earth ended up with a nitrogen-rich atmosphere that contains only trace amounts of carbon dioxide.

The point is that Earth is a unique accumulation of over one hundred elements and element-reacted compounds, and the chance of having two identically formed planets has to be extremely remote. So when coupled with the extremely remote possibility of an Earth, Moon, Jupiter combo, the odds against it happening are, pardon the pun, astronomical.

In the words of Lee Strobel, *"After studying all of the extraordinary circumstances that have contributed to life on Earth, and then overlaying the amazing way in which these conditions also open the door to scientific discoveries, I see design not just in the rarity of life in the universe, but also in the very pattern of habitability and measurability."*

OUR AMAZING MOON

As previously mentioned, our Moon just happens to be exactly the right size and position to support human life on Planet Earth, and I'd like to give it special recognition in this part of my commentary. Is our Moon just a coincidence, or is there more to the story? Does God have something to do with its existence?

I've heard it said that, scientifically speaking, our Moon shouldn't be there. Our Moon is about 196 thousand miles away from us and is currently receding from the Earth at about 1.6 inches per year. This recession is a result of gravitational interaction that's caused by the motion of tides on Earth. When the Moon was closer to the Earth, , the gravitational and tidal effects would have been quite different. Obviously, the recession rate is not a linear calculation of 1.6 inches per year times 4.0 billion years (145 thousand miles). In other words, we can't just multiply the presumed number of years of the Moon's existence by 1.6 inches per year to determine how close it was to the Earth at that time. However, it is mathematically possible to make this determination.

What do you think would happen on Earth if the Moon was orbiting only a few thousand miles away rather than its current 196 thousand miles? Imagine its effect on our gravity, spin rate, atmosphere, and ocean tides. In 1962, MIT professor of geophysics Louis Slichter reported that his calculations indicated that the Moon could not be 4.0 billion years old, because it would have been too close to the Earth at a much earlier time (between 1.4 and 2.3 billion years ago). Because science's evolutionary model requires four billion years, this became a dilemma that could cause major problems for the theory of macro evolution. So, as expected, the evolutionists refined Slichter's calculations to account for Earth's land mass changes over the supposed 4.6-billion-year time frame, and made their calculated results coincide with their theory. The problem was that in order for the new calculations to work, they needed to assume almost no tidal friction at the beginning of the Moon's recession.

In other words, if one assumed that the Earth was completely covered with water and had no land mass at all, the Moon would not move away from the Earth. By strategically locating landmasses at Earth's poles and at the equator there would be some tidal friction to cause the Moon to recede, but at a much slower rate than Slichter's prediction. So by manipulating these landmasses, their calculations were made to show a four- billion-year-old Moon at its current 296 thousand-mile location and having its current measured recession rate of 1.6 inches per year. Strangely however, all of science's early Earth models show a singular landmass running from north to south (known as "Pangaea") and not located in the way that the revised Moon recession model requires. Are we being manipulated by scientists whose aim is to convince us that the Earth is 4.6 billion years old?

A recent Science Channel program explained that a large planet-sized object collided with the Earth about four billion years ago. During the collision, the Earth's gravity pulled the heavier elements away from the Moon debris, thus making the Moon only 60 percent as dense as the Earth. After the collision, the Moon debris coalesced to form a Moon-sized spherical object and travel in an orbit that was about 14 thousand miles from the Earth. Because of tidal interactions, the Moon has since receded to the point where it is now. Using this impact model, the Science Channel once again had a scientific spokesperson confidently and all knowingly pronounce that this is how the Moon was formed and why it is where it is today.

As I see it, the evolution model is bogus in the first place and should not be a premise that requires force-fitting Moon recession calculations to keep it intact. Although I haven't discussed the 4.6 billion-year evolution model, it should be noted that this model was itself force-fitted to account for new scientific information that had developed over the last hundred-plus years. The most important new information was that proteins cannot develop in the presence of oxygen. To accommodate this requirement, the Earth needed to have an oxygen-free atmosphere for single-cell microbes to miraculously form. However, these early microbes needed to inhale carbon dioxide and exhale oxygen to a point where the Earth's atmosphere became what it is today; a process that required about two billion years. As the atmosphere changed, other microbes miraculously evolved to inhale oxygen and exhale carbon dioxide, and you know the rest of the story.

Whatever you believe to be the origin of the Moon, it does have some pretty unusual and amazing characteristics. As mentioned in my "are we alone" commentary, it's exactly the right size, it's located in exactly the right place, and it exists at exactly the right time in human history. It's almost as if it had been deliberately placed there for our benefit. If you were to examine the Moon based upon a preconceived notion of how planetary bodies must have been formed, you would probably be quite surprised. Rather than having been formed with heavier elements coalescing at its core, we find an abundance of refractory elements like titanium on its surface. Moreover, seismic testing on the Moon revealed that it acted more like it was hollow rather than being a homogeneous sphere. I'm not going to go into more detail about the Moon's strangeness because that's not my purpose in writing this commentary. If you are interested, there are numerous articles on the Moon's strangeness and anomalies on the internet that range from what seems absurd to what seems to be quite plausible.

What I found most interesting about studying the Moon was that it sparked my imagination about the far distant future of multi-generational space travel. As you know, I believe mortal humans will one day, in the not too distant future, robotically make huge rotating spacecraft like Epsilon 1 from the materials found in the asteroids. However, what I found interesting about the hollow Moon theory is it might be exactly what Epsilon 1 inhabitants, in a far distant future, might attempt as a first step in terraforming an Earth-like planet. After all, the unique characteristics of planet Earth tell me that it was designed to be like it is. The idea that Earth miraculously formed all by itself and then evolved with life on its surface, through natural events, stretches my sense of logic.

MOTHER EARTH

With further regard to the uniqueness of planet Earth, I did some research regarding the "Mother Earth" theory. If you are of a Darwinian evolutionary mindset, you would probably regard humans, animals, plants and microbes as just being living things that developed without any external guidance. As such, you would not ascribe a special significance to humans, except that their brains have advanced to a higher level. In other words, the human brain has evolved and the idea of a soul doesn't exist. Because you have assumed this evolutionary mindset, you probably don't believe in God.

On the other hand, something is probably telling you that you need an explanation for the unexplainable. Perhaps you believe it can be explained by extraterrestrials with technology that's billions of years more advanced than ours that seeded the Cambrian species and terraformed the Earth 530 million years ago. Then perhaps the Earth's biosphere may itself have evolved into a living organism that has a symbiotic relationship with all living things. Over the billions of so-called evolutionary years needed to produce life on Earth, it may have become a controlling life force, – what some people call mother earth or Gaia. I call it part of God's design.

In 1926, a Russian scientist named Vladimir Vernadsky wrote a now famous book titled *Biosfera*, in which he hypothesized, contrary to Darwin's theory, that geochemical processes evolved first and created the atmospheric environment in which living organisms could emerge. In other words, geochemical and biological processes evolved together in a symbiotic relationship. Vernadsky believed that the cycling of inert chemicals on Earth is influenced by the quality and quantity of living matter, and that living matter, in turn, influences the quality and quantity of inert chemicals being cycled throughout the planet.

The biosphere is a thin envelope that stretches thirty or forty miles from the depths of the oceans to the upper stratosphere, and contains all of the forms of life that exist on Earth. Within this narrow band, living creatures and the Earth's geochemical processes appear to interact to sustain life.

In the 1970s, English scientist, James Lovelock, and an American biologist, Lynn Margulis, expanded upon the Vernadsky hypothesis with the publication of the Gaia hypothesis. The publication points out the regulation of oxygen and methane as prime examples of how the process between life and the geochemical cycle works to maintain a homeostatic climatic regime on Earth. Lovelock asked, since oxygen levels on the planet must be confined to a narrow range, how is it possible that the oxygen level is maintained in such a regulated way? A 1 percent rise in oxygen level would increase the likelihood of a fire by 70 percent, and a 4 percent increase would engulf the entire planet in flames, thus destroying all life forms. Oxygen production is maintained by photosynthesis. The green chloroplasts inside plant cells convert the sun's energy into chemical energy for the plant's nurturance. In these processes, carbon dioxide and water are converted to oxygen. As animals take in the oxygen and exhale carbon dioxide the cycle continues.

While scientists have known for quite a while how the oxygen and carbon dioxide cycles interact, they have been baffled as to why the oxygen levels remain so fixed despite major changes in the sun's output and the kinds and number of living creatures inhabiting the planet. To explain what was happening, Lovelock noted that methane was a biological byproduct produced by bacterial fermentation. Microorganisms living inside ruminant animals, in termites, and in peat bogs produce more than 1,000 million tons of methane per year. Methane migrates to the atmosphere, where it acts as a regulator, both adding and subtracting oxygen from the air. As it reaches the stratosphere, methane oxidizes into carbon dioxide and water vapor. The water vapor then separates into hydrogen and oxygen. While the hydrogen escapes into space, the oxygen descends to Earth. On the other hand, when methane is added to the lower atmosphere, the methane oxidizes and uses up the oxygen. Lovelock points out that *"in the absence of methane production, the oxygen concentration would rise by as much as one percent in as little as twenty-four thousand years: a dangerous change, and on the geological timescale, a far too rapid one."*

By using this example, Lovelock and Margulis claimed that there must be some kind of triggering mechanism creating a biological feedback loop to cause microorganisms to increase or decrease oxygen production. In effect, they are saying this mechanism is scientific proof of a symbiotic relationship between geochemical and biological processes, and evidence for a living Earth biosphere.

I believe this remarkable oxygen-controlling mechanism is more evidence for the existence of God, and not the existence of a mother earth entity. Isn't this what God would have done in creating life on Earth? Doesn't the existence of life on Earth also depend on things happening outside the Earth's biosphere, like the sun's intensity, the Earth's magnetic field, the Moon, and Jupiter? In any event, what I've described is what many in the scientific and political realm have recently come to believe, and it's behind much of what is happening in the environmental movement today. This movement believes that they have a mission to sustain the mother earth entity and prevent humans from destroying it.

From the perspective of many in the Gaian mindset, humans have become a cancer that is destroying mother earth, and anything that upsets the natural order, that has evolved over billions of years, needs to be brought under control. To the Gaian extremists, elimination of or control of human overpopulation is a solution. Although the extremists don't have the upper hand, at least not yet, others are actively conditioning people to join their belief system. It's brainwashing on a global scale.

Their beliefs wouldn't be so bad if the consequences of their ultimate plan wasn't harmful to so many people. In my opinion, it's a global power grab in which a select few may end up controlling the lives of many. The failed Soviet communist system shows clearly the folly of having an elite few ruling over the masses. What the Soviets failed to understand was that in a communal society the human need for freedom and the natural instinct to improve their lives and the lives of their immediate family is obscured.

The catapult issue currently driving the environmental movement is so-called "global warming," which the environmentalists claim is caused by unregulated humans and their gluttonous use of fossil fuels. They believe the Industrial Revolution, with all of the technical benefits and medical achievements, was derived from using carbon dioxide-producing fossil fuels, and must now be curtailed. They believe that rather than growing the world economy so that the poverty-stricken peoples of the world can enjoy the benefits of industrialization, we need to redistribute wealth rather than growing it. The people who are advocating these things think they know the truth and we should be wary of them.

Finally, if a person was looking for proof of God's existence, the Gaian science is a very compelling example. Maintaining Earth's consistent oxygen level and other biofeedback systems should be reason enough to believe a controlling force is actively sustaining life on Earth. If life has existed on Earth for more than a billion years, isn't it possible that feedback systems have been in play for a long period of time, including non-biosphere interactions with the Milky Way Galaxy, the sun, the Moon, and our solar system planets. Doesn't it make more sense to believe that the intelligence of God designed these self-regulating systems, and our loyalties should be directed to a creator God? And, if there is a God, doesn't it also make sense that God created mortal humans with a purpose? Could it be that one human purpose might be to appreciate and care for God's works in a way that improves rather than stifles the quality of life for other humans? Could it be that by restraining human technologies, in an attempt to save the planet, might do more harm done than good?

UFOS

After reading the preceding commentary, do you think I believe in UFOs? Well my answer is, yes, I do believe in UFOs. Are you surprised since all of what I've just said provides evidence that Planet Earth is alone in the universe, and alien beings from another planet don't exist? My belief in UFOs is not what you might think it is. Because my explanation requires a considerable amount of background information, I've decided to wait until you've read the following chapters and then finally tell you my answer in the Epilogue.

CHAPTER 3

IS OVERPOPULATION A PROBLEM?

Whether viewed from a young Earth creationist standpoint or from a Darwinian evolutionary standpoint, I've developed more questions than answers regarding the issue of what is considered to be an overpopulation problem. During my research I couldn't find answers that made much sense to me – especially from the evolutionary perspective.

My reason for looking at the overpopulation issue is because it seems strange to me that humans are just now experiencing an exponential growth rate that apparently has not occurred in the past. Is there an explanation for this? After all, it seems like a short-sighted view of this trend is causing many people to become alarmed. As a result, actions are currently being taken that are contrary to what I believe God intends for us. I believe that God has provided an answer to overpopulation, and it's living in space, and possibly populating our vast universe.

THE ARITHMETIC

Let's start our investigation of population growth by doing some arithmetic. Human population growth during various epochs of Earth's history can be calculated using the well-known formula Re: http://ldolphin.org/popul.html:

$$P_n = 2(C^{n-x+1})(C^x-1)/(C-1)$$

where P = population, n = generations, C = ½ the number of children per family, and x = average number of generations alive.

If x = 3 (the number of generations that the parents are alive) and C = 1 (two children per family) the population growth rate is zero.

With plagues, wars, famines, genocides, and natural diseases that shorten the average number of generations parents are alive and the number of children they may have, the population could go down. For instance, it's estimated that 150 million people died from the plague during the mid-1300s. However, in an agrarian society, families tend to have many children, to help with farming chores, which would cause the population to recover and continue increasing in spite of these disasters.

If a human evolutionary model of one million years is correct, then today's human population should be many times greater than it currently is. But, for some reason, evolutionists begin human population growth at 1000 BC with about 300 million people. This may be supported by other presumed population growth models, but it still represents only a small fraction of what we might expect using the above population growth equation, and does seem suspect. Are we **again** being misled by force-fitting of data?

Let's begin with Earth's current population trend. Using the evolutionists' population model, the Earth's population grew slowly from 1000 BC to AD 1750. In this 2,750- year time period, the population grew from about 300 million to about 800 million, in spite of plagues and other population-reduction events. Using the above population growth rate equation and assuming an average life span of 35 years, 3 children per couple, 35 years per generation and beginning with only one couple, my calculations show a population of 4.24 trillion in just 60 generations, and 6,540 trillion in 2,750 years. Since this obviously did not happen, my assumptions in this calculation must obviously be wrong.

Between 1750 and 1950, the population grew to 2.5 billion. Using the same assumptions, but this time starting with 400 million couples (800 million people), my calculations show 9.1 billion people by the year 1950. Again my assumptions in this calculation must obviously be wrong, since they don't include World Wars I and II, the US Civil War, and other world conflicts. Therefore, in spite of these conflicts, the assumed basis for my calculations is close to being right.

By 1985 the population was five billion, and by 2000 it was more than six billion, and future forecasts predict that by 2050 there will be 9.2 billion people and by 2090 there will be 18 billion. If I use the population growth formula and my previous assumptions, but starting with 1.25 billion couples (2.5 billion people) in 1950, my calculations show 8.44 billion people by the year 2050 and 12.65 billion by the year 2090. Now things are starting to make sense. In case you're interested, by the year 2200 my calculations show a population of 42.7 billion and 144.2 billion by the year 2300. This is obviously a problem that needs to be addressed now. Let's get ready by developing nuclear fusion energy, portable hydrogen fuel from water, and moving into space!

The most intriguing thing about the above analysis is the slow population growth during the 2,750-time interval between 1,000 BC and AD 1750. How can this be? While localized famines, wars, and disease could slow down population growth, human extinction certainly didn't occur. Since my calculated population growth rate is obviously overstated, my assumptions do coincide with what we are seeing today. Besides, even if we assumed devastating periodic downward adjustments, we would also have periodic upward adjustments. However, a problem occurs when the upward adjustments start with a new base. In other words, if we start with two people who grow to one million people over a period of time and then have a downward adjustment to only one thousand male and female couples, our new upward adjustment starting point is two thousand rather than two. So if we have a similar upward adjustment timeframe, our end point population would be a million times a million, or one trillion people. Do you see the problem here? In case you are thinking that an agrarian human population will only reach a certain size because of food resources, I agree. But again we are talking about the problematic 2750 timeframe when Earth's food (cropland, animals, and fish) resources could easily sustain at least three billion or more people, even with ancient farming techniques (we have that many people living in poverty today).

The flat human population growth rate over the past three thousand years is quite strange indeed. Even though today's exponential growth rate correlates with the Industrial Revolution, it doesn't make sense to tie the two events together in a cause-and-effect fashion. The lion's share of current population growth is coming from Third World agrarian countries, and not from the industrialized world. Having eight or more children per agrarian family is not uncommon. If so, why didn't the pre-industrial agrarian population grow just as rapidly as it is today, in spite of the aforementioned disasters? Perhaps I am missing something here! Although we no longer have plagues, we still have wars, famines, genocides, and naturally occurring disasters. Yet facts are facts, and I can only ask why. Is God intervening here?

Currently, about twenty-five thousand people die of starvation each day, mostly children under the age of five. Yet as you can see, this hasn't slowed our exploding population growth much, if at all. Does this mean starvation kept the population from exploding in the past? If so, where is the historical evidence? Moreover, scientists claim about twenty-seven thousand square miles of the world's cropland is currently being lost each year because of a lack of water and soil degradation. In spite of this, it hasn't slowed today's population growth very much. Fortunately, current technology has been able to keep the food problem from getting out of control. These technologies include increased crop yields, drip watering, soil rejuvenation, electric pump access to underground aquifers, modern-day transportation, and refrigeration. Without these technological advances, today's population problem would be much worse. If today's population had existed in the past, it could have been disastrous. Could it be that God has intervened in the past to prevent extreme poverty and starvation? Is God intervening now to show us how to solve our population problem, so we can fulfill a greater plan? As you might imagine, our current population trend is quite disturbing and will probably cause an unsustainable demand for Earth's non-renewable natural resources. This trend will cause an increased interest in mining our solar system. Moreover, it will probably lead to my prediction about the use of helium 3 fuel and multi-generational space travel.

HOW ABOUT ANIMALS?

When considering the implications of the population growth equation, it's surprising we are not overrun with many animal species today. After all, it takes only one male and female animal to begin the process. If we go back to the surviving species after the presumed extinction of dinosaurs sixty-five million years ago, there would have been enough time to overwhelm the Earth with the animals that were not prey to other animals. My internet research on this subject provided sketchy answers. I'm certainly not an expert, but when I asked my internet questions I found evasive answers with suspect science backing up claims. When scientific spokespersons talk about limited food resources, a mysterious balance of predator and prey, and procreation self-control when populations became overcrowded, it doesn't fit with my sense of logic! Perhaps it's instinct that limits animal procreation, but what uncontrolled evolutionary process would cause it? And if humans are evolved animals, why don't we possess this same instinct with our currently increasing population growth?

It's been stated the Great Plains buffalo reached sixty million by the mid 1800s, why not much more than this number if they have been around for millions of evolutionary years? Did the buffalo population peak at sixty million when the prairie grass could only sustain this many? If so, how did this peak population occur? Was it their instinct to stop procreating or did the buffalos have a meeting to discuss the subject and establish reproduction rules? If limited food was the cause of limiting population growth, why wouldn't the buffalo migrate to find more food? After all, North America is a big place!

If the buffalo have been around for millions of years without human intervention to cull the herd, the aforementioned population formula should apply. My research indicates a wild buffalo's life span ranges from 15 to 20 years, and females have an average of 10 to 15 offspring over their lifetimes. I guess I am just confused. Although today's human population is expanding out of control, for some unknown reason the animal population has been kept in check. Could it be, in cases where animal populations had become too large that events, human or otherwise, have mysteriously occurred to make this possible?

Imagine a herd of sixty million buffalo that roamed the Great Plains in the mid 1800s. In the movie *Dances with Wolves*, the incredible sight and sound of a stampeding herd of buffalo was portrayed in a way that has left a lasting impression on me. Although the ground shook like an earthquake and the thunderous sound of millions of hoofs hitting the ground was deafening, the most profound effect on me was seeing the vast swath of grassland that was trampled after the animals had passed. Today the buffalo still roam in areas like Yellowstone National Park, and the Black Hills of South Dakota, but not in the great numbers of the past. Perhaps the demise of the buffalo at the hands of humans actually served a greater good, since their prior Midwest habitat is now the breadbasket that is helping to feed the United States and the world. If the buffalo had not been hunted to near extinction, their numbers might have exceeded 120 million today, and the land area would be uninhabitable for farm families. Is this just mere coincidence?

I don't want to belabor the point about the potential problem of an uncontrolled animal population, but it does seem strange that events caused by humans or otherwise have maintained a semblance of animal population control. Think about the trillions of locusts that devoured the vegetation in the same Midwest buffalo habitat, but have themselves mysteriously disappeared. Perhaps if humans had not hunted the buffalo to near extinction, the locusts might have caused them to starve. Perhaps locusts were responsible for controlling the buffalo herd in a continuous cycle of renewal over millions of years. Consider a world overrun by any animal, especially some undesirable animals, birds, snakes, or insects! What is preventing this from happening? A better question might be, is the millions of years' assumption wrong?

EXTINCTION

Why would an animal species that has survived hundreds of millions of years go extinct within the last few thousand years? And, if for some reason they do go extinct, are there any significant consequences? After all, the only cataclysmic event that has happened in the last few thousand years is the advent of humans. Even if humans deprived animals of their habitats (like cutting down the rainforest), micro evolution (adaption) should help them to survive. Because of my questioning, I looked up animal extinctions on the internet. To my surprise and amazement, there were hundreds of websites. Many of these websites implied that humans are like weeds currently killing off thousands of animal species, and that this will have catastrophic effects because we are upsetting nature's ecosystem. Since scientists claim that 95 percent of all previously living species have gone extinct before the advent of humans, the Earth seems to have done pretty well in spite of this presumed fact.

Many of the website bloggers claiming scientific evidence for animal extinctions also claim these facts alone prove the young Earth creationists are absolutely wrong. Their arrogance goes so far as to use derogatory terms to describe the young Earth creationists, like *loony*, *psychotic*, and *delusional*. While name-calling is not in the best interest of scientific discourse, these website bloggers do seem adamant about forcing creationists to become extinct. What I find interesting is that all of their claimed scientific anti-young Earth arguments can also be counter-argued. Moreover, the sheer number of these self-reinforcing websites tells me that the authors, are not just annoyed by the young Earth creationists, they are also panicked by them. Rather than welcoming the creationist's arguments as a means to strengthen their own science, they seem to be protecting an agenda that transcends the science.

While I am not claiming to be a young Earth creationist, I have at least studied and evaluated what they have to say. In contrast it seems like the supporters of an old Earth and Darwinian evolution are acting like the child who covers his ears and makes loud noises to prevent hearing anything that is being said by his parents. Is this what we have come to?

GOD'S PLAN

Before leaving the subject of human and animal population growth, I'd like to explain a previously mentioned observation, but in more detail. No matter how large the human population becomes, there appears to have always been a way to deal with it. Although many alarmists see our population growth as an insurmountable problem, I see it as part of God's plan. While

I've already discussed human migration to other solar systems as the ultimate solution to our increasing population, there would need to be a transitional solution, since a tripling of our human population may occur within the next hundred years. If you believe that God is behind all that is happening, then you might consider that fossil fuels may have been the first transitional source of energy that allowed the Industrial Revolution to occur. If so, then there would need to be at least one or more transitional steps since fossil fuels are non-renewable and presumably in limited supply. My guess is hydrogen made from an almost unlimited supply of water will serve as our next portable power source, and nuclear fusion will be our next stationary power source.

Miraculously we can use hydrogen, made from water, as an alternative transportation fuel. Even at predicted future fuel consumption rates, the amount of water needed to make hydrogen is miniscule compared to the amount of water available on Earth. Is this just a coincidence, or does God have something to do with it? Conventional nuclear fusion electric power (not derived from helium 3) will probably be commercially viable by the year 2050. Surprisingly the fuel for this power would come from an element contained in water called *deuterium* and a substance called *tritium*, which is derived from the Earth's abundant supply of lithium. Therefore, from a technical feasibility standpoint, we will not run out of fuel before we are technically capable of "mining the sky," which I will discuss later.

On the other hand, we will begin stretching our ability to supply construction materials, water, and food. Although fresh water may become an issue in the future, grand canals could assure the required supply, as discussed at length in my book,

Reaching America's Destiny. My guess is that drip watering of crops, hydroponics, recycling, and large-scale zeppelins may provide some answers to our materials and food problem. Do you see the picture that's developing here? Does God have anything to do with these coincidences, or is it just another miracle of human ingenuity?

While writing about the population problem, it occurred to me that some people may have religious convictions about where mortal humans are headed in the future. For more than two thousand years, many believers have thought that Christ's return is imminent. While I don't discount the idea that this may be true, I believe when Christ returns it will be much further into the future, a future in which most mortal humans have acknowledged Christ's existence and are living in the sky in Epsilon 1-type satellites. At that time, Christ could more easily appear and communicate with all of the then existing humans.

When I think about the population issue, it makes me believe that Earth's mortal humans are unique in the universe. If another more advanced civilization had existed somewhere else, God would probably have figured out the human population problem. Are Earth's humans the first prototypes in need of refinement before the final design is put on the market? Or is there another explanation? The Flood and other events described in the Bible indicate that God has made what appear to be mistakes in the past, but is capable of making course corrections as needed. Some of these apparent mistakes include changing mortal human life spans after the Flood, preventing pre-Flood hybrid offspring from becoming part of the human family, and not anticipating how long it would take for post-Flood humans to acquire the technology needed to sustain an exponentially growing population. Hopefully, with our current population problem, we will develop transitional technologies and access our solar system materials, before it becomes too late for our survival.

CHAPTER 4

IS MAN CAUSING CLIMATE CHANGE?

Years ago there was a popular television program called *Dragnet*. It involved two police detectives named Sergeant Joe Friday and his sidekick Officer Bill Gannon who investigated various crimes. A famous line Joe Friday would say is, *"Give me the facts ma'am, just the facts."* So if you would like to debate what I'm about to present, I suggest you come prepared with facts and not unfounded opinions. Political correctness, name-calling, and irrelevant information won't do it for me.

SEEKING THE TRUTH

Let me begin by saying we need critics. We need educated critics because they point out things that might have been overlooked when a plan of action has been formulated. However, I'm concerned how critics often have a hidden political agenda. As a consequence, some good ideas get scuttled before they see the light of day. In the previously stated words of Mark Twain, *"I admire those who seek the truth, but I am wary of those who claim to have found it."* Everything I've presented in this book is based on science, mathematical facts, and logic as I understand them to be. I don't have a political agenda. If anyone reading this book can show me where my facts and understanding are wrong, I'm more than willing to listen and debate them. If your point of view is convincing enough, I'll gladly change or modify mine.

I need to be cautious here because many people claim to have found the truth in their religious beliefs. Fortunately, many of these believers are continually looking for proof to validate their premise. In other words, they examine facts with an open mind. I consider myself to be this kind of religious believer. If scientific, archeological, or other evidence proved that the Bible is incorrect, it would cast doubt on what I believe. If historical or scientific information indicates other possible explanations, I would examine them further. If extraterrestrials were to make themselves known, I would want to find out what they believe and why (including how they came into existence in the first place). However, I haven't yet found anything that would change my religious beliefs. In fact, everything I've researched has caused my religious beliefs to become even stronger.

My problem is with people who have been taught what to think rather than how to think. When young people are taught what to think, we experience Hitler Youth, Kamikaze pilots, lynch mobs, and followers of Jim Jones. Are you easily led, or do you think for yourself? Do you question what you believe? Have you been taught that something is a scientific fact, without challenging the evidence? With today's internet, almost anything can be checked, examined, and verified. Look for counter-arguments to see if there's any validity in what others are saying. Open your mind to different points of view. Try to recognize if you have become indoctrinated or hoodwinked. Remember Mark Twain's quote, *"It's easier to fool people than for people to recognize that they have been fooled."* Don't be fooled by those who have an ulterior motive, like advancing a political agenda.

Many times decisions are made before doing simple arithmetic. When someone claims there is a hundred years' supply of oil, a simple calculation using a 2 percent per year compounded growth model could prove the fallacy of this statement. That's not to say some gigantic oil reservoir might be discovered in the future, but without proof, it's just wishful thinking and not fact. This is especially true with regard to energy and energy policy. When our energy policies are based upon a supposed fear of manmade global warming, we are being totally unrealistic, especially when the underdeveloped world is building thousands of coal fired power plants, and we're trying to solve the supposed problem by driving hybrid cars and taxing our carbon emissions. Then again is there a hidden political agenda behind driving hybrid cars and taxing carbon emissions? I think there is, and here's why.

To again quote renowned genius and Manhattan Project scientist Richard Feynman, *"Science is a belief in the ignorance of the experts."* So whilst there are many scientists who sincerely believe that manmade carbon dioxide is causing catastrophic global warming, keep in mind that much of what they believe is based upon half-truths and incomplete scientific knowledge; especially a belief in unverified computer models that are claimed to be capable of predicting the effects of carbon dioxide on the climate's future! In fact, they may be using a few scientific facts to keep their incomes flowing from government study grants. Since politicians around the world appear to believe the current politically correct idea of catastrophic manmade global warming, they have a tendency to require a litmus test before they grant research money.

IS IT REALLY JUST POLITICS?

Could it be that manmade global warming is really a political issue that has nothing to do with preventing the oceans from rising? Let's review some quotes by some prominent people and organizations:

"The common enemy of humanity is man. In searching for a new enemy to unite us, we came up with the idea that pollution, the threat of global warming, water shortages, famine, and the like would fit the bill. All these dangers are caused by human intervention, and it is only through changed attitudes and behavior that they can be overcome. The real enemy then is humanity itself."

— CLUB OF ROME
Premier environmental think tank consultants to the
United Nations

"We've got to ride this global warming issue. Even if the theory of global warming is wrong, we will be doing the right thing in terms of economic and environmental policy."

— TIMOTHY WIRTH
President of the UN Foundation

"The models are convenient fictions that provide something very useful."

DR. DAVID FRAME
— Climate modeler, Oxford University

"It doesn't matter what is true, it only matters what people believe is true."

— PAUL WATSON
Co-founder of Greenpeace

"We are on the verge of a global transformation. All we need is the right major crisis."

— DAVID ROCKEFELLER
Club of Rome executive member

"My three main goals would be to reduce human population to about 100 million worldwide, destroy the industrial infrastructure and see wilderness, with its full complement of species, returning throughout the world."

— DAVE FOREMAN
Co-founder of Earth First

"A total population of 250–300 million people, a 95 percent decline from present levels would be ideal."

— TED TURNER
Founder of CNN and major donor to the UN

"The Earth has cancer and the cancer is man."

— CLUB OF ROME
Mankind at the Turning Point

If you would like to read more quotes like these, please visit the website www.green-agenda.com. Authors of other similar quotes include Barack Obama, Tony Blair, Al Gore, Prince Philip, Mikhail Gorbachev, Harry Reid, Gordon Brown, and Angela Merkel. While their intentions may actually be good, I'm concerned that their utopian ideas are much too radical for me to accept. Culling the Earth's human population through legislation as a solution to global warming, water shortages, and other concerns is unimaginable. Yet here are prominent people and organizations, using global warming scare tactics, to help create a one world government (a world without boarders – a la John Lenin's song "Imagine," that, when in power, *may* resort to extreme measures. Do these extreme measures include sterilization of people deemed unsuitable to have children? Do these measures include mass exterminations of undesirables? Does food deprivation provide a convenient population reduction solution? Do they intend for most of the world's population to live a meager agrarian existence, while only a select few members of the ruling class enjoy the benefits of our modern industrial age? Will population growth be curtailed by enforced legislation of one child maximum per family?

In any event, it has the ring of Hitler and Nazi Germany. If Hitler had prevailed, and had absolute power, just imagine what he would have done to solve the population issue. A clear and unequivocal example of how absolute control results in misery and hardships is the failed Soviet Union. If you are one who believes that under the "right" leadership, a Soviet style of government can work, you may be right, but as the old saying goes, "absolute power corrupts absolutely." Without checks and balances, who knows whether the person in charge will be a smart and benevolent dictator or a selfish and corrupt tyrant?

Even if the prominent people supporting the global warming agenda do not harbor the idea of implementing extreme measures, they might not have thought it through to a logical conclusion. After all, the idea of redistributing wealth sounds commendable when you see the disparity between the rich and poor, and the squalor and despair so many of the world's population must endure. But, enriching the impoverished by impoverishing the rich just doesn't make sense. If you tax the rich too much, they will lose their incentive to innovate and create jobs. As a result, the redistributed wealth will be so diluted everyone will be impoverished; except, of course, for the ruling class. This is typical of what can be called *static thinking*. It's thinking that doesn't account for a dynamic reality like the fact that the rich will lose their incentive and economic growth will be stifled. Besides, it's fallacious to believe in a zero-sum game. Wealth can and should be expanded so there will be enough for everyone. And, the most tried and true wealth expansion, or growth, process is free markets. It's what has made the United States the economic powerhouse it is today, and it's why China's economy is currently growing so fast. Fortunately, China's current dictatorial government understands the benefits of a free market system; and it's my guess that eventually the Chinese people will demand their freedom, and move to a more just form of government.

By claiming global warming is manmade, the quoted enthusiasts are in effect brainwashing the public. After the brainwashing is completed, they probably expect the public to accept the redistribution of global wealth through carbon tax legislation and similar measures. They might also accept higher energy costs and a resulting lower standard of living in order to "save the planet." Included in this brainwashing will be a belief that "green energy," produced by wind turbines and solar energy, will replace the nasty carbon dioxide-producing fossil fuels and dangerous nuclear power. Well folks, these ideas are a house of cards. With regard to green energy, I will describe in the next chapter some possible solutions to using wind and solar energy more effectively, so I am not suggesting that we throw the baby out with the bathwater. There are better ways to transition away from fossil fuels than to tax carbon dioxide and over-regulate electric power companies that use fossil or nuclear fuels.

THE CRITICS

Consider this: more than thirty-one thousand American scientists have signed the Global Warming Petition Project. These scientists include more than nine thousand PhDs, and the signers include world-renowned physicists such as Prof. Edward Teller and Prof. Freeman Dyson. Nearly four thousand of the signers are scientists trained in specialties directly related to the physical environment of the Earth, and the past and current phenomena that effect the environment.

The petition states:

"There is no convincing scientific evidence that human release of carbon dioxide, methane, or other greenhouse gasses is causing or will, in the foreseeable future, cause catastrophic heating of the Earth's atmosphere and disrupt the Earth's climate. Moreover, there is substantial scientific evidence that increases in atmospheric carbon dioxide produce material beneficial effects upon the plant and animal environments of the Earth."

Perhaps even after what I've presented so far, you're still convinced that manmade catastrophic global warming is real, and that wind turbines will save the day, let me present some additional facts, even though these facts may cause you to experience cognitive dissonance.

AN INCONVENIENT TRUTH

I'd like to begin by explaining what I believe to be the fundamental reason why global warming and global cooling exist, and why it's a naturally occurring phenomena. Due to a combination of axial precession and tilt, and varying orbital changes around the sun, resulting from gravitational interaction with other planets in our solar system, Earth is expected to experience an extreme freezing and thawing effect about every twenty thousand years. Earth's precession cycle is about twenty-one thousand years. Its axis tilt angle relative to the sun is about forty-one thousand years. The orbital eccentricity cycle is about one hundred thousand years, and the orbital plane angle is about seventy thousand years. These phenomena are called Milankovich cycles. What this means is that during a twenty-thousand-year cycle, Earth will experience varying degrees of warming and cooling because of its relationship to the sun.

In fact, recent history shows a mini ice age occurred from AD 1300 to 1500. At that time people living in London, England were ice skating on the Thames River. Before that time, we had a medieval warming period from AD 800 to 1300. This was a time when Viking explorers settled on Greenland's warm and hospitable shores. Before that time, the Earth experienced a "dark age," cold period, between AD 300 and 800.

Needless to say, manmade carbon dioxide was not the culprit in causing these periods of global heating and cooling. In fact, if you have been influenced by the famous "hockey stick" graph showing a correlation between rising global temperatures and recent industrial age carbon dioxide emissions, you might want to reexamine your conclusions. Since the "hockey stick" graph did not accurately display past global heating and cooling periods, it has been discredited. In fact, it has been ridiculed by the widely viewed and famous "You Tube" video skit called *Hide the Decline.*

Of course the cause of global heating and cooling is not just attributable to Milankovich cycles. If it were not for water vapor, carbon dioxide, and other atmospheric gases, the Earth's global temperature would be much lower than it is. However, increasing the amount of water vapor and/or carbon dioxide doesn't equate to a linear increase in global temperatures. In other words, by doubling the amount of carbon dioxide, its effect on global temperatures is not a doubling of its temperature influence when acting by itself. Because of this interaction between water vapor and carbon dioxide the influence of currently increasing atmospheric carbon dioxide levels has almost no consequence.

Rather than go into a lot of detail showing the fallacy of catastrophic manmade global warming, I recommend that you visit this website, www.theresilientEarth.com. When people read the articles provided in this website, they will see the possibility that manmade global warming is a manmade hoax.

Of all the things I've read regarding the fallacy of manmade global warming, the thing I found most intriguing is a finding made by a prominent climate scientist and advisor to the United Nations named Susan Solomon. Susan found that from 2000 to 2009 water vapor in the stratosphere had diminished, with the result of depressing global warming by about 25 percent compared to that which could be attributed to carbon dioxide and other greenhouse gasses. In other words, as atmospheric carbon dioxide levels were rising, the Earth somehow accommodated this rise by reducing its stratospheric water vapor content. Wow! This is indeed strange. Is God intervening here? Also – consider this. As we all know, when global temperatures rise, the oceans evaporate and clouds form. Increased cloud formation means increased precipitation and snow fall. Between the increased snowfall and the clouds, more sunlight is reflected back into space. As a result, the Earth's temperature will tend to fall. Somehow the Earth appears to be protecting itself from a "runaway" temperature effect.

THE GLOBAL WARMING HOAX

Further, I recently read a book entitled, *The Greatest Hoax*, by United States Senator David Inhofe. It was a real page-turner and eye–opener. I could hardly put it down. The book centered on the global warming debate in the senate, but what made it so fascinating to me was that it described many interesting aspects about the inner workings of our government.

In his book, Inhofe explains that he had become the lone outspoken skeptic of manmade global warming in our Congress. He began his skepticism by questioning the predicted devastating consequences of man's contribution to global warming. His main argument against legislation like "cap and trade" was that it would destroy the US economy by imposing the biggest tax increase in history. Inhofe stated the people who would be hurt the most were the poor, because energy expenses are a bigger portion of their income. Not only would it cost taxpayers about $4 hundred billion per year, it would also cause many US manufactures to move production facilities to countries like China, India, and Brazil. Since these countries don't have our stringent pollution controls or plans to reduce carbon dioxide emissions, pollution and carbon dioxide emissions would probably go up. Thus, without worldwide participation in the carbon dioxide reduction program, the US legislation would be "*all pain and no gain.*" It was Inhofe's argument of "*all pain and no gain*" that has so far kept the United States Congress from passing cap and trade into law.

In 2003 there were few in Congress who doubted the dire consequences of manmade global warming. Although there were some skeptics, only Senator Inhofe was outspoken about it. Doing this during the peak of the global warming frenzy was tantamount to announcing he believed the Earth was flat. After all, the United Nations panel of renowned scientists wrote reports saying manmade global warming was real. This was the gold standard and who could doubt it? The problem was the actual report did not definitively state that global warming was associated with increasing manmade carbon dioxide emissions, but a politically motivated executive summary did. Unfortunately, Congress and the media only read the executive summary. Ultimately, the Climategate scandal discredited the United Nations' reporting, especially with the fudged data behind the famous hockey stick graph, which showed the twentieth century to be the warmest in the last two thousand years.

I'm not going to go into the numerous examples of discredited science behind the global warming scare, but politically motivated science can work both ways. While I have the highest regard for Senator Inhofe's courage and perseverance, I would caution him about information provided in his book about the availability of fossil fuels in the United

States. Even though the United Nations' scientific reports were falsely reported by politically motivated executive summaries, Senator Inhofe's statements about the availability of fossil fuels in the United States, are somewhat misleading and lack a global perspective with regard to worldwide demand for oil.

On pages 164 and 165, in Inhofe's book, he states: *"the United States is endowed with 163 billion barrels of recoverable oil. That's enough oil to maintain America's current rate of consumption and replace imports from the Persian Gulf for more than fifty years."* In my opinion this statement is misleading because it doesn't use the words *"undiscovered technically recoverable."* In other words, it is a scientifically based educated guess. Until the oil is actually discovered and is shown to be economically recoverable, we shouldn't base US policy upon the hope that it will become a reality. Besides, 163 billion barrels of oil is not all that much when viewed globally; but, it can help to free the United States from imported oil in a relatively short period of time. As such, I applaud Senator Inhofe for advocating that we minimize obstructing regulations and make federal land and offshore regions available so we can explore and drill for the "undiscovered" oil as soon as possible. After all, there is a clear and present danger to our economic viability and national security and an aggressive drilling program will help immensely. And, in my opinion, so will starting an aggressive program that phases in hydrogen. Let's do both!

Let's face it, eventually we are going to run out of fossil fuels if we continue on our current path. Therefore, doesn't it make sense to develop alternate ways of making energy? While I have no problem with energy research, I do have a problem with researching ways of removing carbon dioxide from smokestack emissions. Currently, the most favored way is called *"carbon sequestration."* The most common sequestration scheme is to pump the carbon dioxide gas to more than a mile underground. This is an expensive process that if done on a massive scale would be impractical. Imagine pumping thousands of cubic miles of carbon dioxide into the ground; it just doesn't make sense. I've done the arithmetic. Yet experts talk about it as if it can and should be done. Where is our common sense, especially if these actions will do nothing to prevent naturally occurring global warming or cooling?

When science programs on television tell us that humans are spewing trillions of pounds of carbon dioxide into the atmosphere each year, it sounds ominous, until you again do some simple arithmetic. When you do the arithmetic, you find that these trillions of pounds are miniscule when compared to the volume and weight of the Earth's atmosphere. As one scientist puts it, if you doubled the amount of airborne carbon dioxide in the Earth's atmosphere, it would be like adding one more person to a crowd of 70 thousand at a Sunday afternoon football game. Moreover, as previously mentioned, a doubling of manmade carbon dioxide does almost nothing to affect an increase in the Earth's temperature.

With regard to manmade carbon dioxide and other really harmful pollutants, consider this: when transportation energy is derived only from hydrogen, and electric power is derived only from nuclear fuel and renewable sources, the global warming issue will go away. However, as you now know, global warming isn't the issue at all; it is overpopulation and a political attempt to form a one-world government. So even if the carbon dioxide issue were solved, it wouldn't satisfy the zealots. This means that my hydrogen fuel recommendation will meet with additional claims of disaster in spite of it being the only practical solution to the supposed manmade global warming issue.

In conclusion, the Earth will experience global temperature fluctuations regardless of human activity, Milankovich cycles. While increased amounts of carbon dioxide are currently being emitted into our atmosphere, its effect is negligible. By transitioning to hydrogen as a portable fuel and nuclear fusion as a stationary fuel, the political issue will go away. Let's be rational, and less provocative, and speed up this process so that no one is embarrassed by illogical and politically motivated suppositions. Sometimes a common-sense solution is needed to bring people together.

CHAPTER 5

DO WE NEED
RENEWABLE ENERGY?

There were these two brothers. One was a pessimist and the other was an optimist, and both wanted a pony for Christmas. On Christmas morning they went down stairs and found a room full of horse manure. The pessimist brother expressed disappointment and said, "I didn't think we would get a pony." The optimist brother said. "There must be a pony in here somewhere."

LOOKING FOR A PONY

Using the optimist/pessimist story as an opening to what I'm about to say, I'm going to begin my discussion about renewable energy in an unusual way. From an engineering perspective, the technology of wind and solar energy is fascinating. It reminds me of when Westinghouse moved me from their Heat Transfer Division in Philadelphia to their Nuclear Steam Generator Division in Tampa, Florida in 1969. At that time, I maintained contact with some of my friends in Philadelphia. During one telephone conversation I was told that Westinghouse's Heat Transfer Division had been given a government contract to develop an Ocean Thermal Energy Conversion (OTEC) system.

After investigating, I found that OTEC had two to eight percent energy conversion efficiency and used very expensive materials that required more energy to make than the system would provide in its lifetime. At that point I called my friend and asked why their engineers would work on something so obviously impractical. His answer really caught me off guard. He said that he and the other engineers working on the project were aware of the impracticality, but the engineering challenge was a lot of fun and they liked working on it. Since that time, it occurred to me that OTEC could be made very practical if combined with the waste heat generated by modular navy nuclear power So the point I'm making is that engineers are quite clever at turning lemons into lemonade; and this holds true with regard various forms of wind and solar energy. Their optimism makes them think that there's a pony in there somewhere.

Well, what I'm about to say reflects on my optimistic viewpoint. While much of today's renewable energy looks like a boondoggle when fossil fuel and nuclear power is much more efficient and reliable, I tend to believe there must be a pony in there somewhere.

As most of you know, the term *Imagineering*, was invented by the Walt Disney Corporation, and I want to give full credit to them for it. Having spent most of my life in the engineering profession, the term *Imagineering* is inspirational to me. I have been particularly inspired by this quote by George Bernard Shaw and later made famous by John F. Kennedy, *"Some people see things the way they are and say why. I dream of things that never were and say why not?"* The word *Imagineering* sums up this quote in one word.

It's hard to explain how fortunate I feel to have worked in the engineering profession, and how proud I am to have been associated with the remarkable contributions to humanity that my fellow engineers and scientists have made. Engineers not only invent solutions to problems, they also tend to make seemingly endless incremental changes that make the invention better and better. Anyone who has driven a modern day automobile knows what I'm talking about. Moreover, it goes without saying that the driving force behind these changes is free enterprise competition in the marketplace.

Engineers love to solve problems through invention and technological innovation. To my mind, problem solving can range from designing an automobile to fixing an economy, and everything in between. Just provide the known facts, and engineers will systematically reach a sensible conclusion -- too bad more of our politicians aren't engineers! In some cases, a new invention is needed to reach the next level. Apple's I phone and I pad come to mind as I'm trying to think of a good example. During the I phone's prototype stage, it was discovered that keys in a person's pocket could scratch the screen surface. As the story goes, Apple's founder Steve Jobs contacted the CEO of the Corning Glass Company, Wendell Weeks, and asked him to develop a glass that was scratch-resistant. Weeks recalled a material that was developed for a mothballed project from the 1960s, and the rest is history. It's now called "Gorilla Glass."

So how, you may ask, do engineers, and even non-engineering people, keep coming up with new ideas? First of all, in my humble opinion, you need to believe that God has a solution for every problem and your job is to look for clues that God has provided. However, I am well aware that many of my fellow engineers don't see it that way. Anyway, finding the clues is another matter entirely. The best way I can describe how to find the clues is to recall a quote from Louis Pasteur, *"In the fields of observation, chance favors only the prepared mind."* In my case, I inadvertently stumble upon clues, because I've geared my mind to solving specific problems. As an example, I may watch a Science Channel program on television, or read a science book in which a scientist says something that causes me to see something that hadn't yet occurred to me. From this glimmer of new knowledge, I begin to fit more pieces of the problem puzzle together. Finally, after the problem has been formulated I sleep on it. Amazingly, the next morning an answer to the problem often becomes evident to me. This process involves a mysterious aspect of the subconscious mind, which may be the way in which God has created humans. Perhaps the subconscious mind is somehow connected to universal knowledge.

If God is providing clues for us to solve our problems, then almost everything we see around us needs to be examined from this perspective. With fossil fuels being a limited resource, the problem of eventually finding a replacement becomes obvious. Realizing nuclear power is one obvious replacement for stationary electric power, a replacement for portable power is also needed. Since hydrogen made from water can be that replacement, it seems obvious that our vast supply of water provides a solution. Considering the potential need to provide our growing population with portable fuel, the water–to–hydrogen solution seems to have been given to us as part of God's plan for us humans. Is the fact that water can be converted into an almost unlimited source of portable fuel just another fortuitous coincidence, or is there a deliberate reason? Was the fact that fossil fuels were available to drive our industrial revolution, with its advanced technologies, just a fortunate coincidence, or was it made available to us for a reason?

With the above random thoughts in mind, let's now discuss renewable energy.

RENEWABLE ENERGY

In its current form, renewable energy doesn't make sense; yet, if viewed in a different way, it could make sense. While wind and solar energy are free, the apparatus needed to extract it is generally too expensive and occupies too much land area. Moreover, the wind and solar energy is too unreliable. Additionally, wind turbines kill birds and create a lot of noise. Besides, with fission and (eventually) fusion energy being available, who needs them?

What do I mean by viewing renewable energy in a different way? How about making hydrogen fuel rather than electrical energy that must be consumed when produced? In this way, we can produce the hydrogen fuel whenever the wind blows or the sun shines. The hydrogen can be stored and delivered without regard for the demand side of the equation. To make this idea a reality, we should consider huge interconnected energy parks in remote parts of the country that compensate for sunshine and wind variances. Onsite mobile mass-production facilities could be implemented to minimize construction and transportation costs; and by combining two or more forms of energy production we could minimize land space requirements. Moreover, solar collectors that capture solar heat during the day and store it to make electricity at night makes sense to me.

Since we would be locating these energy producing facilities/ parks in remote unpopulated regions of the United States, we could construct long distance superconducting transmission lines to populated regions to minimize energy transportation costs and energy losses. Then, what if we manufacture equipment that lasts forty or more years, so investments are spread out over a long time frame before needing replacement (current wind and solar equipment life expectancy is twenty-five years).

Then how about artificially creating ocean thermal gradients using waste heat from modular navy nuclear reactors (my previously mentioned invention described in my book *Reaching America's Destiny*)? This proposed OTEC system not only makes the nuclear power systems more efficient, it also creates a substantial amount of fresh water as a byproduct of making electricity. If we then use the fresh water to make hydrogen, the need for long underwater electric transmission lines can be replaced by liquid hydrogen tanker ships that can deliver hydrogen fuel to populated coastal states. Moreover, how about replacing the tanker ships with zeppelins? By the way, one more benefit of the OTEC system is that it brings deep water nutrients to the ocean surface, which allows sea life to thrive and promote fishing.

AN ENERGY OVERVIEW

In case the previous suggestions are providing too much information too fast, let's slow down and describe the various forms of energy production. I'll start by asking these questions: Is there enough land to produce biofuels, wind power, and solar power and still feed a growing world population? Is there enough uranium to produce nuclear fission power? What about geothermal and hydroelectric power? Will nuclear fusion power ever become a reality? Can we derive power from our oceans? Is space going to provide the answer?

To answer these provocative questions, I'll begin with a brief description of the various energy-producing options so that misconceptions can be put aside.

Hydro and Geothermal

You may be asking why not use hydroelectric, and/or geothermal to produce hydrogen fuel and/or electric power? Actually, we are using hydroelectric and geothermal power, but the US and world's energy consumption needs are so great that available hydroelectric and conventional geothermal sites fall far short of consumption requirements.

Currently, non-conventional versus conventional geothermal energy production is being given serious consideration. Conventional geothermal uses steam from reservoirs close to the Earth's surface, and can be easily tapped to produce electric power. These sites are generally beneath scenic treasures like Yellowstone National Park or on Native American reservations.

Non-conventional geothermal means extracting the heat energy that's found in dry rock about three miles or more underground. To do this, water must be pumped into the ground such that it fractures the hot rock and redistributes itself to a number of adjacent extraction wells. If this resource could be tapped, just 2 percent of the available underground heat in the United States could provide nearly two thousand times the power that the nation now consumes each year. As it is with extracting energy from the sun or the wind, the problem is cost. Current estimates range from between a competitive 10.5 cents per kilowatt-hour to a sky-high $1.05 per kilowatt-hour. However, perhaps my proposed hybrid energy park could use the wind turbine pedestals to double as geothermal drilling rigs and reduce the cost of both energy producing systems.

Wind Power

Wind power can be generated by many forms, including air rotors, autogiros, ocean tides, and ocean waves (all of which are described in my book *Reaching America's Destiny*), but the most commonly known form is the wind turbine. I'll discuss only this form because it appears to have the most potential for becoming competitive with conventional power-generating systems.

Not too long ago, my wife visited Pipestone, Minnesota, where she picked up a brochure from the US Department of the Interior regarding wind turbines located in the state. When I read the brochure it sounded impressive, touting the fact that eleven hundred $2.5 million, 1.6 megawatt wind turbines had combined to produce two million kilowatt-hours of electricity per year. However, when I did some simple arithmetic, I found that these turbines produced an average of only 228.3 kilowatts per hour of intermittent power, which is only 13.83 percent of their rated capacity. To put this into perspective, a conventional fossil-fired or nuclear power plant can produce dependable, on demand, electricity for about one third of the wind turbine life cycle cost, even though the wind is free. As if their cost and intermittent power output weren't enough of a problem, these wind turbines occupied about one hundred fifty times the land area normally required for conventional electric power generation. Moreover, in addition to their visual blight, they create a disturbing noise when operating in high winds and have a propensity for killing birds, including endangered birds like the golden eagle.

Wind turbine costs have decreased significantly over the past two decades, and will continue to drop, albeit at a lesser rate, with technological innovation and increased volume. Additionally, the idea of distributed wind farms is currently becoming popular, since if they are interconnected in a grid fashion, they will tend to offset the lack of wind from one location to another. However, there is a glitch that no one seems to be addressing in a logical way. The glitch is that the wind turbines need to be very large in order to be more efficient and cost effective. Because of their large size, they pose significant factory–to-site logistics problems. To give you an idea of the logistics problems, consider this: Minnesota's eleven hundred 1.65 megawatt $2.5 million wind turbines have a pedestal base diameter of 16 feet, run 30 feet into the ground, and weigh 2.9 million pounds. Imagine delivering these pedestals and their enormous blades from a remote factory to the installation site. Now imagine dealing with a much larger 5 megawatt wind turbine with blades that are 206 feet long!

To further exacerbate the issue, land usage can become a huge problem if the United States intends to produce a large percentage of electricity using wind turbines. If only 30 percent of our electricity was derived from 1.65 megawatt wind turbines, they would occupy a land area equivalent to almost all of North and South Dakota. If 5 megawatt turbines were used, they would occupy only the land equivalent of almost all of the State of North Dakota. I chose using these states for a reason, since according to a "wind map" of the United States, most of the Class 4 to 5 annual winds are located in the High Plains corridor adjacent to the eastern side of the Rocky Mountains. This area is not only far removed from high energy use regions of the country, but much of it sits on top of the Ogallala Aquifer (to be discussed later).

Because of their huge land area requirements, high costs, intermittent and unreliable wind conditions, and high electric transmission losses, it's obvious that we need to find a single solution that addresses all of these issues. In my opinion we need to look outside the box, while at the same time, considering other energy alternatives that will result in an overall comprehensive energy strategy in which all of the pieces of the energy puzzle fit together.

To address the land use and cost issues, related to wind turbines, I propose building three large (sixty miles long by sixty miles wide) remotely located energy parks that utilize renewable, and possibly non-renewable, energy systems. Because mobile tent factories could be located on each site, the cost issues related to wind turbine transportation would be minimized and a form of mass production could be devised. Moreover, since other proposed energy systems are integrated into my proposed idea, the land area requirement and cost of each piggybacked energy system could be significantly reduced, while at the same time the issue of intermittent and unreliable wind conditions could be ameliorated.

Although my proposal may sound like an answer, the problem of electric transmission line losses would become a major issue in the economic equation. To reduce the electric transmission line losses, I propose using superconductivity (electric power transmitted without line losses when the transmission wires are at a very low temperature). ﹖

As an additional superconductivity benefit, it was shown in one instance, that eighteen thousand pounds of copper wire could be replaced by only two hundred fifty pounds of superconducting wire. The problem is that the cost of currently proposed cooling systems outweighs the benefit.

On th e other hand, high temperature superconductivity has been demonstrated in the laboratory and when ready for commercialization, in about five years, it could be a game-changer.

Solar Power

Solar energy producing systems can take many forms, including heliostats, parabolic thermal collectors, sterling engines, ocean thermal gradients, and space-based satellites that beam electric energy to Earth via microwaves (all of which are described in my book *Reaching America's Destiny*); however, most people are familiar with photovoltaic (PV) solar panels. Although Southern California Edison has constructed more than 1,000 megawatts (peak rated output) of solar panel electricity, I believe this form of electric power needs further development before it can compete with conventional forms of large-scale power generation. I say this because the typical solar panel, using low cost thin film technology, is capable of converting only about 13 to 17 percent of the sun's energy into electricity, and during a typical day and night only about 17 percent of this energy is available. In other words, if we had a square mile of photovoltaic solar panels operating in a southern desert we would get about 42 megawatts of power during a sunny day, but over a 24-hour period, only about 6.3 megawatts of power will be generated (these numbers account for energy reductions due to sunrise and sunset, dust accumulation, and proper spacing).

While the capital cost for photovoltaic solar panels has dropped dramatically over recent years from about $25 per watt to about $1 per watt, the installed life cycle cost is still about four times more than conventional power generation, even when conventional power fuel costs are taken into account. On the other hand, thin film silicon-based photovoltaic solar panels appear to be on the verge of a breakthrough. This breakthrough uses nanotechnology that allows sunlight to become trapped in three versus two dimensions. As a result, about 25 to 30 percent of the sun's rays could be captured rather than 13 to 17 percent. According to my arithmetic, if 30 percent efficiency could be obtained and the price were to drop to $0.50 per watt, we would be at a breakeven point with conventional power. However, a 1,000 megawatt solar plant would still occupy about seventy square miles of land (about nine miles long by eight miles wide) versus about 1.2 square miles for a conventional 1,000 megawatt coal, natural gas, or nuclear plant.

Nuclear Power

The recent earthquake and tsunami in Japan have left many people believing that nuclear power is no longer a viable energy option. I disagree. Nuclear power is not only where we should be going, it is where we must go! I don't want to minimize the consequences and hardships caused by the Fukushima nuclear reactors, but we need to put things into perspective.

Before the earthquake, Japan had fifty-five water-cooled reactors that generated about 30 percent of its electricity, and they are all located in earthquake-prone zones. Although the reactors at the Fukushima site were affected, the other reactors survived without damage. And, as you may know by now, the Fukushima reactors have been brought under control with no known loss of life. With regard to radiation that was released into the atmosphere, the concern has been greatly exaggerated. Ted Rockwell, a noted Manhattan Project scientist, has stated that people around the world live in radiation levels that are much higher than is present in the Fukushima evacuation zone without showing any ill effects.

Remember, the Fukushima accident was caused by an unprecedented magnitude 9.0 earthquake that was later followed by a major tsunami. When looked at as a whole, the earthquake and tsunami caused much more loss of life and property than the Fukushima plant. Damage from exploding oil tanks, burst gas mains, electrical fires, and other forms of devastation were much more harmful to the Japanese population.

Although engineers had designed the Fukushima reactors to withstand large-magnitude earthquakes, and other events such as the impact of a large jet airliner, they were apparently not designed to lose backup electric power when inundated by a tsunami. When flooding caused backup diesel generators to become inoperable, the pumped water, needed to cool the reactor cores, could not be provided. However, due to the successful implementation of extraordinary measures, the reactors and spent fuel rods were cooled and the resulting damage was minimal, especially when compared to the Russian Chernobyl disaster.

I'm not going to go into all of the redundant safety aspects of well-designed modern-day nuclear power plants, but I can assure you from my firsthand knowledge, as an engineer who worked alongside Westinghouse's nuclear scientists and engineers, the chance of a major Chernobyl-like accident is statistically impossible. However, since the extremely unlikely Japanese incident did happen, we will learn from it and incorporate appropriate backup safety systems in present and future fission-type nuclear plants.

With the above being said, the four hundred thirty-eight nuclear power plants currently operating around the world provide about 17 percent of the world's electric power. Another thirty plants are under construction and one hundred four more plants are in the planning stage. Of the total number of operating nuclear plants, one hundred three are in the United States and provide about 20 percent of our electric power. Nuclear power plants have an almost flawless forty-year track record, and we shouldn't throw the baby out with the bathwater. However, we should be aware that fissionable nuclear fuel is not an unlimited resource. At the current worldwide consumption rate, the world's known uranium reserves will only provide about fifty-two more years of fuel. Moreover, if all of the world's electric power were derived from currently designed fission nuclear plants our known uranium reserves would be depleted in about nine years. So what can be done?

Because uranium is a limited and non-renewable resource, a different type of reactor is needed to extend the uranium fuel supply. It's called a "breeder" reactor. Although the breeder reactor can extend the duration of uranium usage by about sixty times, a different style of breeder reactor, called the "Very High Temperature Reactor" (VHTR), can also extend uranium usage, but only about thirteen times. Instead of generating electricity, the VHTR can also be used to economically produce hydrogen fuel from water using what is called the sulfur-iodine thermo-chemical process, which is explained in my book *HYDROGEN*. While many people remain fearful of having too many fission-type nuclear reactors, I ask that you consider them as a bridge to a future that utilizes only nuclear fusion.

Nuclear fusion is mostly derived from water; it does not create radioactive waste and it cannot experience a meltdown;

but it's still in an experimental stage. In my opinion, it may require twenty-five to fifty more years before it becomes a viable energy source. While fusion derived from water may be the ultimate energy source, I believe that it is just another bridge to the next step, helium 3.

As previously mentioned, helium 3 can be found on the Moon and in the upper atmospheres of the gaseous planets and doesn't occur naturally on planet Earth. Nuclear scientists will readily agree that helium 3 is the best fuel for a fusion reaction, and has the potential for direct conversion to electricity. If so, it may one day be possible to miniaturize the process to where it could fuel an automobile or provide, from one tank full, enough electricity for the life expectancy of a home.

SUMMARIZING

In summary, hybrid wind and solar power could be used to economically produce a significant quantity of hydrogen from water by implementing large-scale energy parks in remote parts of the United States and delivering it to populated areas via underground superconducting transmission lines, pipelines, or remote-controlled zeppelins. Although not yet mentioned, for economic and practical reasons, I recommend using 12 megawatt wind turbines rather than the current maximum size of 5 megawatts. Also with the advent of low-cost/high-efficiency 3D photovoltaic panels and advanced modular parabolic solar thermal systems America could provide a very useful form of foreign aid to developing countries. More about this later.

CHAPTER 6

IS THERE A
WATER SHORTAGE PROBLEM?

I f you have ever stood next to a beautiful lake, or the ocean, you know we humans have an affinity for water. Perhaps, since 98 percent of a human being is water, there appears to be a natural tendency for them to be drawn to it. To help prove this point, there is a lake near Minneapolis, Minnesota, called Lake Minnetonka. It's about fourteen miles long by about three miles wide, but it's made up of many mini-lakes, rivers, and secluded ponds. Since it's close to a large city, people can live on the shores of this lake and commute to their workplace. Because of this convenience, many of the homes built on or near the lake are big and expensive, and it's getting to the point where only the wealthy can afford to live there. In fact, I've been told that one stretch of homes overlooking the lake, comprise the most expensive real estate in the country. Names such as Dayton and Pillsbury live there.

My wife and I have a summer home in Minnesota, a state claiming there are ten thousand lakes, plus Lake Superior. From my observation there appear to be many more lakes and large ponds than the claim. For the last few years we have vacationed with our daughter and her family on the north shore of Lake

Superior. If you've never been there, the beauty is absolutely magnificent and breathtaking. After growing up near the Atlantic Ocean, I can tell you that Lake Superior looks like an ocean. But I will say that it definitely has a different feeling to it, like no salt air.

During one summer vacation we took the grandkids on a Lake Superior chartered fishing trip to catch lake trout and salmon. To catch these fish, the fishing boat needed to troll at a pretty fast speed. However, since the fish were swimming near the bottom of the lake, special outriggers with very heavy weights were required to bring our lines down to where the fish were. To my surprise the lake was over two hundred feet deep at less than a quarter mile from shore. It was at this point that I recalled my brother-in-law telling me of his concern about a worldwide water shortage. But with all of this fresh water how could that be? As an engineer who believes technology can overcome almost any physical obstacle, I thought there must be a way to use this vast amount of fresh water without emptying it into the Atlantic Ocean by way of Niagara Falls and the St. Lawrence Seaway. My thinking is that if fresh water ever became enough of a crisis that my fellow engineers would solve the problem.

FRESH WATER

When I studied my brother-in-law's water shortage concern, I found he was right. A shortage of fresh water could become a crisis in many parts of the world. As the population of humans and animals increases, the need for more fresh water is becoming quite evident. The problem of course is not that there isn't enough fresh water, but rather that there's not enough of it in the right places. According to the United

States Geological Survey (USGS) there are 2,548,339 cubic miles of fresh water located in rivers, lakes, and ground water throughout the Earth – almost all of it being in underground water aquifers. In addition, there is another 3,088,000 cubic miles of saline groundwater.

The problem with our naturally occurring hydrologic cycle is that it doesn't distribute water evenly throughout the globe. Some areas of the globe have an abundance of water, like Lake Superior, while other areas don't. One article I read claimed that six countries have 50 percent of the Earth's fresh water. These countries were identified as Brazil, Russia, Canada, Indonesia, China, and Columbia. Unfortunately, one-third of the world's population is located in what might be called "water stressed" countries.

Although diverting fresh water by using dams is a prevalent method of getting water to where it's needed, it does create some problems. Over 60 percent of the world's rivers have been dammed, with over forty-seven thousand dams being built in just the last fifty seventy years. The problems resulting from this damming process include downstream erosion due to sediment buildup at the dam site; and in many cases land near the dam site becomes flooded and fish migration routes become blocked.

While the lack of fresh water in some regions of the globe may sound ominous, let's look at some facts. First of all, worldwide human water consumption varies greatly. For example, in Ethiopia the average person consumes only about three gallons of water per day. In Great Britain, the water consumption is about thirty gallons per day, while in the United States it's between seventy and one hundred fifty gallons per day. In other words, we in the United States could continue our existence on much less water if we had to do it. Additionally, in California, where a water shortage is claimed, the pricing of water has been so low that the first eight hundred eighty-five gallons, costs about $2.80. After that, the price increases to about $3.40. Since the average Northern California family of three consumes about three hundred fifty gallons of water per day, their bill is only about $35 per month. Perhaps if they had to pay $95 per month they would think about taking shorter showers and conserving water. If so, the California water shortage problem would probably go away.

In any event, maybe it wouldn't be such a bad idea to charge more for fresh water because it isn't like charging more for gasoline. A higher price for gasoline, and diesel fuel has a detrimental effect on the economy because for many people and businesses cutting back is not an option; whereas raising the price of water causes most people to use it more wisely. Perhaps if water were to cost more, private enterprise and the free market might come up with novel new ways of extracting water from water-rich regions and delivering it to water-poor regions.

The idea that we are running out of water is most certainly a myth when you consider desalinization. Current methods of desalinization can produce tens of millions of gallons of water per day for about $0.015 per gallon (depending on the cost of energy). For a person using one hundred fifty gallons per day, the cost would be $2.30 per day, or less than $70.00 per month. Since the water needs to be transported to the location where it's used, the cost will of course go up and investments and profits need to be added. In one case a desalination plant in Jubail, Saudi Arabia transports water via a two-hundred-mile pipeline to the city of Riyadh. Of course the problem with desalinization is the availability and cost of power needed to achieve it.

When water transportation costs become prohibitive, especially to highly elevated inland regions like Mexico City, it becomes cheaper to transport water from a fresh water source and eliminate the capital cost and energy consumption associated with desalinization. The problem lies with poor inland locations that can't afford to import water. So it isn't that there's a lack of water; it's that poor people can't afford it.

THE OGALLALA AQUIFER AND COACHELLA VALLEY

At this point, I want to discuss the Ogallala Aquifer and California's Coachella Valley since – as you will soon find out – they can play an important role in my proposal to help solve the water issue in the United States.

THE OGALLALA AQUIFER

The Ogallala Aquifer is located under portions of eight Midwest states, and is one of the world's largest underground reservoirs and lies beneath 174 thousand square miles of land, located to the east of the Rocky Mountains, which is often called the "bread basket of the United States." However, this was not always the case. Have you ever heard about the semi-arid High Plains crop failures in the 1930s, due to cycles of draught, culminating in the disastrous "Dust Bowl?" This event preceded the development of electric pumps to extract water from the aquifer.

The water depth of the Ogallala ranges between one hundred to four hundred feet, and is between three to five hundred twenty-five feet below the high plains surface, and is comprised of about 970 trillion gallons. Although this is certainly a lot of water, farming in this area is consuming the Ogallala's water at ten times the rate of replenishment. This is due to evaporation by the High Plains arid atmosphere, the impermeability of the top layer of soil (called *caliche*), and small pore spaces in the underground natural resupply network. This problem is not unique to the Ogallala; it generally exists for other aquifers throughout the world. Although current estimates state the Ogallala is being depleted by as much as 5.7 trillion gallons per year, there might be enough water left for one hundred seventy more years. However, the depletion is not uniform, and some farming regions are going to run out in as few as twenty-five years if nothing is done to reverse the depletion trend.

THE COACHELLA VALLEY

Another much smaller aquifer exists under much of the Mojave and Sonora Deserts, but steps are currently being taken to replenish its depleted water via a canal connected to the Colorado River. This canal also supplies water to residents of the Coachella Valley. The one-thousand-square mile Coachella Valley is located mostly in Riverside County in Southern California and boarders on Imperial and San Diego Counties. Since this part of the country receives only about three inches of rain per year, it needs to utilize water from the canal. The Coachella area is home to more than 250 thousand residents and grows high-value crops on one hundred ten square miles of land. In fact, these crops have a gross annual value of $575 million, or about $8,000 per acre, which is among the highest crop dollar returns per acre in the world.

The one-hundred-twenty-two-mile long canal, supplied by the Colorado River, provides water for crops, replenishment of aquifer water, and for general use by the Coachella population. Water sales to the population are about 40.3 billion gallons per year, but 6.5 billion gallons of wastewater is reclaimed for watering crops, lawns, and golf courses. To conserve water usage for crops, drip watering and micro irrigation is extensively employed with focused applications of fertilizer, pesticides, and herbicides being distributed using these same systems. My estimate is that about 40 billion gallons is needed to grow crops in the 110 square mile area. This is equivalent to about two feet of rainwater in one year, but because the water is only needed during the growing seasons, this amount is concentrated to grow crops at that time. With this in mind, the Coachella Valley is part of my proposed economic model for future water management and I will refer to it later as we proceed.

AN ECONOMIC INCENTIVE

I mention the Ogallala Aquifer and the Coachella Valley because, in the plan I'll propose, we can build a grand canal in the United States that will not only replenish and sustain the Ogallala, but can also bring additional water to the Mojave and Sonora Deserts and allow more Coachella-like valleys to be developed. If my plan proves to be economically viable, it can be duplicated in other parts of the world and help eliminate future water and food shortages. Because the project will be massive, and expensive, the trick will be to provide economic incentives.

Let's begin by assuming seventy additional one hundred ten-square mile Coachella Valley crop-growing areas in the Mojave and Sonora Deserts. This will result in a gross yearly income of about $8,000 per acre, or about $40 billion per year.

In addition, wouldn't the land value increase significantly as people, especially retirees, move into this area for the climate and scenery? Now, let's assume that 2.8 trillion gallons of water would be needed to make this happen, and that it can be sold for $0.004 per gallon. This would result in sales of about $11.2 billion per year. Going further, let's assume that 10,000 square miles of this improved land, or 420 million acres, could be sold for the modest price of $10,000 per acre, the profit would be $4.2 trillion.

AMERICA'S GRAND CANAL

Now I'll explain how I propose building "America's Grand Canal," and how it can be funded. Since Lake Superior has experienced high water levels of about two feet during a spring thaw, the amount of water contained in these two feet is nearly 13 trillion gallons. By building a large pipeline/aqueduct/canal system from Lake Superior to the Mojave Desert, we can pick up additional water from the Missouri, Red, and Arkansas Rivers (and also control spring flooding from these rivers).

The northern leg of the proposed pipeline/aqueduct could be comprised of two twenty-five-foot diameter tunnels from Lake Superior to the Missouri River. This leg of the canal could be pumped uphill, below the frost line, to the Missouri River using the English Channel "Chunnel" method. A system of concrete pipes, aqueducts, rivers, and holding lakes could then be used to transport water from the Missouri River to the Mojave Desert.

Now let's do some arithmetic to determine what this project might cost. I'll assume that construction of the Grand Canal will require thirty thousand workers per year over a twenty-year timeframe. This equates to about $4 billion per year for labor at $50 per hour. I'll now be generous and assume that another $6 billion is required per year for land, equipment, materials, and energy. This equates to $10 billion per year, or a total expenditure of $200 billion in twenty years. Now assume that half of the canal can be constructed in ten years and supply about 80 percent of the water needed for a Coachella Valley project. At this point, we can begin selling land and water.

Over the second ten years of construction we might sell half of the supplied water and 5 percent of the land. Assuming this is a gradual process, this ten-year sale of land and water could equal about $266 billion, or more than the cost of building the canal. During the next twenty years, forty total years, we could sell water and the remaining land for about $250 billion per year. After forty years the income would be derived from water sales alone, about $11.2 billion per year.

Based upon the above analysis, I propose that the Grand Canal be funded by the government and private investors. Although it could be funded by private investors alone, I see a need for the government to be involved. Here's why. Besides making government land and resources available, the entire nation benefits from saving the Ogallala and preventing river overflow flooding. Preserving the Ogallala could be considered a national defense issue, and preventing river overflow flooding would not only save lives and property, it would also save emergency relief money. And, from a long-term point of view, it will provide revenue for America's social safety nets. The plan also creates long-term, high-paying jobs for thousands of people, and tax revenue income from food that's exported from the newly productive cropland.

So as you can see, building an American Grand Canal can not only provide water to water-stressed regions of the country, it can also provide significant economic benefits that could serve as a demonstration to the rest of the world. Please refer to my book *Reaching America's Destiny* for more details regarding this idea, including large hydrogen-producing energy parks, superconducting electric transmission lines, and building huge new Midwest cities.

Now let's discuss how we can *save* water by transitioning to hydroponic farming.

HYDROPONIC FARMING

Did you know that hydroponic farms can save 70 to 90 percent of water needed to grow crops? Did you know that hydroponic farms may have a ten to fifteen-fold increase in crop yield and not require pesticides? When considering the water shortage issue, the idea of using much less water to grow food has a definite appeal. So I looked into the subject and here's what I found.

When I read a book about vertical hydroponic farming. I was somewhat disappointed because of a lack of design details and economic justification. Instead, I found a lot of irrelevant opinions that had little to do with what I wanted to know. The book advocated building a few tall glass-enclosed buildings in the inner cities to provide locally produced organically grown food. It also advocated having people move from the suburbs to the inner city to take advantage of public transportation, cultural activities, and other benefits. With this urban idea in mind, it was thought that people would consume less gasoline and reduce greenhouse gas emissions. My thinking is just the opposite. People need to move out of the congested, crime-ridden cities, and the further out the better.

Anyone who can do simple arithmetic could show the folly of inner-city vertical farms. For instance, if a city were to have ten million inhabitants, how many of these tall vertical farm buildings would actually be required to feed them? Using the book's example of growing five crops of corn per year with a yield of about fourteen ears per square foot, a three-hundred square foot space would yield 4,200 ears of corn. That equivalent amount of food will feed about one person for a year. So for the purpose of continuing the arithmetic let's assume three-hundred square feet is required per year to feed one person. If so, then we would need 10 million times 300 square feet to feed 10 million people. To further the arithmetic, let's assume a square building that's about 250 feet on each side. If an inside circular hydroponic crop garden of two hundred twenty-five feet in diameter were constructed, it would equal about one acre of crops. One acre of hydroponic crops might feed about 145 people at three-hundred square feet per person. Since we need to feed ten million people we would need about seventy thousand crop layers. Being generous in saying the layers could be spaced ten feet high, we would need to construct our building to a height of 700,000 feet. Assuming a skyscraper limit of one thousand feet, we would need to construct seven hundred of these gigantic buildings.

As a point of reference I chose the former World Trade Center buildings, once located at the south end of New York's Manhattan Island. Each square building was two hundred eight feet on each side and rose to a height of 1,362 feet. The cost of constructing the twin towers was about $2.3 billion in today's dollars, and they occupied sixteen acres of land. While our seven hundred hydroponic crop-growing buildings may or may not cost $1.15 billion each, it could be close to this number. Using sixteen acres per two towers we would need 5,600 acres (8.75 square miles) of prime real estate for our seven hundred buildings. Since the five boroughs of New York City are comprised of three hundred three square miles, it's hard to imagine how 8.75 square miles can be made economically available for our seven hundred skyscrapers. And then how do we accommodate more people coming into the city at the same time? If we used a practical height of one hundred feet per building, we would need 10 times 8.75 square miles. Folks, it just ain't going to happen. On the other hand, perhaps the inner-city idea should be abandoned in favor of other vertical, or just single-story hydroponic farm possibilities.

While the concept of growing hydroponic crops vertically may sound good at first, there is a glitch: glass-enclosed buildings are expensive, and the energy and equipment required to produce artificial lighting and temperature control could overwhelm any derived benefits. To give an example, light emitting diode (LED) lighting required for 2.0 acres of three layers of vertically stacked hydroponic crops could cost about $1.7 million at today's prices of $2 per watt. The cost of a three-layer reinforced concrete and insulated glass building would add another $1.4 million, including labor. If these costs are amortized over 15 years at 5 percent, the yearly payment would be about $305,000. To make matters worse, the energy needed to turn the overhead LED lights on for an average of eighteen hours per day for one year would cost about $330,000 at $0.1 per kilowatt-hour. The economic problem worsens when you consider the LED lights would need to be replaced every fifteen years. And then there is the cost of farm supplies, heating, dehumidifying, auxiliary equipment, labor, and maintenance, which need to be taken into account.

Making Hydroponics Work

Not to be deterred, I began looking at the vertical farm in a different way. After all, what if the scientist who discovered Preparation H had stopped at Preparation G! Although this may not be an appropriate analogy, I think it makes a point: we need to be persistent and look outside the box.

So, being persistent, let's build a two-level building and forget costly LED lighting. Let's make it with a lower level, which can be used for a variety of purposes, such as manufacturing and food processing. Now let's assume that 30 percent efficient vertically placed photovoltaic solar panels (at $1.00 per watt) are incorporated into the design and make them like venetian blinds that track the sun and are adjustable to a near horizontal position during high winds or sand storms. Moreover, if we add hydrogen electrolysis equipment, 60 percent efficient hydrogen fuel cells, an 80-foot-high by 1,000-foot solar array could generate 107 kilowatts of continuous on demand electric power (assuming a semi-tropical location).

Using the above assumptions, a top level two-acre hydroponic farm consisting of a 1,000-foot long by 100-foot wide concrete and glass building (including solar/hydrogen/fuel cell electric power) could be built for about $1.85 million or about $175 thousand per year using a 15 year/5 percent amortized loan. Now let's continue with our arithmetic and see how much gross profit could be generated. Using sweet corn at 14 ears per square foot per year we could produce 1.22 million ears. I'll now assume fresh sweet corn sells for about $0.50 per ear at the grocery store or farmers' market, and we can get a wholesale price of $0.20 per ear. If we multiply these numbers, the resulting gross income is about $244,000 per year. After subtracting the $175,000 mortgage, we get $69,000 in gross profit per year (without accounting for labor, seeds, fertilizer, auxiliary equipment, water, sewage disposal, trash pickup, insurance, taxes, and maintenance expenses). However, now we're getting close to making the idea work.

Now let's assume that a single family owns and operates this facility and rents out 50,000 square feet of ground-floor factory space (half of the available space) plus about half of the available electricity for about $150,000 per year. With the second half of the available ground-floor factory space, an owner family could process the hydroponic food (canning and freezing) and sell it for a higher price. If my assumed costs are correct, things are looking pretty good at this point. Also, consider the added income that will result when the 15-year amortized loan is paid off.

Considering the analysis, hydroponically grown crops could be the wave of the future, and here's why. With as much as five times greater crop yields (using only sunlight rather than LED lighting) the amount of land required for growing hydroponic crops would be much less. Water usage would be greatly reduced and weeding would be unnecessary. Pesticides would be eliminated, and transportation and farming equipment, with its associated fuel cost, could be minimized. Instead of nearby towns and cities paying to dump grey water from sewage treatment plants into nearby rivers and streams, how about using it to fertilize the hydroponic crops and paying the hydroponic farmer to do it? We could also use manure from farm animals. Fish ponds are also a possibility. Please refer to my book *Reaching America's Destiny* for details regarding design specifics.

Looking Ahead

If the hydroponic farm concept were to be applied to the Ogallala Aquifer region the 5 trillion gallons per year aquifer depletion rate could be reduced to about 1.5 trillion gallons. In addition, less land would be required, and higher valued crops normally produced in California could be grown in a mid-west region with resulting lower transportation costs for delivery to the east coast.

Here is the result of my calculations:

- One square mile of land can accommodate sixty, two-acre hydroponic crop farm buildings and livestock.

- One square mile of hydroponic farms could generate as much as 170 million kilowatt-hours of on demand energy per year. Additional energy could be generated using crop residue, cellulostic ethanol.

- One square mile of hydroponic farms could feed approximately 17,000 people.

- One farmer, with family help, could maintain one two-acre farm, associated livestock, and food products, including processing/packaging (approximate net income = $100,000 per year per farm family).

- The ground floor of the hydroponic buildings could provide space for manufacturing facilities and provide rental income.

Hydroponic Investment Profit

Let's look at what I have just proposed in terms of making a profit for investors. With my estimates in mind, let's assume an investor constructs one two-acre hydroponic farm building each year and makes a 10 percent return on investment, but uses that money to reduce the cost of succeeding hydroponic farm building investments. At the end of eleven years there is a breakeven point, and at the end of twenty-one years the original investment would be paid back with properties valued at $38.85 million (assuming no compound interest or inflation). If we project out to the fiftieth year, the income per year would be $7.215 million and the properties would be valued at $92.5 million.

If the proposed hydroponic farm complexes were located near a populated area, several additional advantages could result. For instance, using returnable food containers rather than boxes and cans might make sense. In the "old days" processed fruits and vegetables were put in sealed jars that were later cleaned and reused. With regard to manufactured products I can think of many alternatives to cardboard boxes that could be retuned or recycled. Obviously the amount of fuel required for conventional distribution would be greatly reduced.

I like the idea of towns with 20,000 to 25,000 people, because it's the number of people in the town of Hastings, Nebraska, where my former employer, Thermo King, has a factory. Twenty-five thousand people can justify two golf courses, many restaurants, a shopping mall, a medical center/ hospital, a small airport, a Wal-Mart, several churches, and a few hotels. Much of the food for this population size can probably be provided by adjacent two square mile hydroponic farm complexes.

Feeding the World -- Hydroponically

Let's expand our minds a little further. Here is where my hydroponic farm idea gets really interesting. Let's use these hydroponic farms to help 2 billion destitute people around the world who are trying to survive on an average family income of about two dollars per day ($730 per year). Is it possible that hydroponic farming could be a way to provide them with the material and medical benefits currently enjoyed by the industrialized world? Let's see:

Let's begin by imagining 1,000-foot long by 100-foot-wide concrete and glass buildings with PV panels mounted vertically on posts. On the ground floor of the building we could create thirty 30-foot wide by 20-foot long three bedroom and one bath, air-conditioned condominiums along the south facing side with adjacent 10 ten-foot-wide side entrances. On the north facing side, we have thirty water tanks that contain grey water, possibly produced by fish. In between the condominiums and water tanks we have a space that can be used in a variety of ways, such as processing food, housing animals, manufacturing new products, or recycling worn-out products like automobiles. Each condominium has an adjacent 30-foot wide interior and exterior space to be used for a variety of purposes, such as maintaining a horse and cart. If thirty families with an average of five people lived in each condominium, the hydroponically produced food could feed each family and the solar-generated electric power would be more than sufficient to supply their needs plus some left over for manufacturing.

If we were to arrange the hydroponic/manufacturing combos in a checkerboard fashion, one two-mile by two-mile square could comfortably house 2,160 families or about 10,800 people. At the corners of the "checkerboard" squares are sewage and waste disposal facilities, hydrogen production and fuel cell electric power generation, a reservoir and water supply system, and an area to make concrete and grind/melt scrap materials. In the center portion of each square we could have a zeppelin transportation system, a runway for small aircraft, and a village center. The village center could include schools, churches, a medical facility, a mini Wal-Mart, a McDonalds, a recreation center, a movie theater an auditorium, a market, a bank, an office building, etcetera. The checkerboard pattern allows for replication and expansion, but I've simplified it here to illustrate what might be possible. Many other more attractive versions of the checkerboard pattern may be possible; however, the buildings would need to be lined up and face in a southerly direction to maximize the use of solar power. In case you're wondering, the land area required for two billion people is approximately 4.5 per cent of the habitable land area of the world. More on this later.

THE BIG PICTURE

In case you're wondering how the last three chapters fit in with my premise that science is discovering God, I'll begin by quoting Albert Einstein who once said *"There are two ways to live your life. One is though nothing is a miracle. The other is though everything is a miracle."* Obviously, I subscribe to the latter case. But, many of those subscribing to the former case, look at today's exponentially growing human population, so called manmade global warming, dangerous nuclear power, polluting fossil fuels, water shortages, and food shortages as circumstances that need to be solved by establishing a one world government. A government that will redistribute wealth, control population growth, employ only renewable energy derived electric power, limit water usage, and distribute food and medical care through a process of central planning. As a result, everyone will live in poverty – except for the elite ruling class – and human freedom will be discarded.

Because, I believe that God will miraculously accommodate exponential population growth, and the other aforementioned circumstances, I see a totally different picture. In my opinion, God has miraculously provided the wherewithal to solve our apparent problems and mankind will employ its recently discovered miraculous technologies to assure our freedom and allow *all* of God's children to prosper. Hydrogen fuel, nuclear fusion, grand canals, global recycling, hydroponics, and eventually living in space, will allow this to happen. As a result of this future process, and evolving scientific discoveries, I believe most of mankind will discover God and the importance of Jesus Christ in their lives.

With this thought in mind, let's examine some additional things that have led me to this conclusion.

CHAPTER 7

WAS THERE
A WORLDWIDE FLOOD?

I n his book, *In the Beginning*, MIT physicist Dr. Walt Brown theorizes that before the global flood there was a subterranean ocean about ten miles beneath Earth's crust. Brown's theory is based upon the Bible's Genesis account. Genesis Chapter 7 Verse 11 states, *"In the six hundredth year of Noah's life in the second month, on the seventeenth day of the month, on the same day all the **fountains of the great deep** burst open, and the floodgates of the sky were opened."* [emphasis mine]

I can only speculate on how the subterranean water formed, but consider this. If the Earth was molten at one time, there would be a sequence for the formation and distribution of Earth's elements and compounds as the Earth cooled. Probably the most prevalent process would be the oxidation of elements like iron. As a result, the lighter weight iron oxides would rise to the surface, and the heavier non-oxidizing elements, like gold, would sink toward Earth's center. In this process oxygen and hydrogen would form water molecules when the temperature reached about five thousand degrees Fahrenheit. To me it's conceivable that water could get trapped ten miles below the Earth's surface as the lightweight oxygen and hydrogen atoms tried to escape into the atmosphere during the solidification process.

THE KOLA SUPERDEEP BOREHOLE

I'm not a geologist, but here is what I discovered when I researched the subject of subterranean water. In 1962 Soviet researchers began drilling the deepest hole in Earth's upper crust – the Kola Superdeep Borehole. By 1994 they stopped drilling at a depth of 7.5 miles, which was 1.7 miles short of their intended goal. Contrary to scientific opinion at that time, they didn't find the transition from granite to basalt between two to four miles of depth that was predicted by seismic testing. Even more surprising was the *discovery of water*. Another unexpected find was a "menagerie" of fossils as deep as 4.2 miles and despite the harsh environment of heat and pressure their microscopic remains were remarkably intact. The researchers were also surprised at how quickly the temperatures rose as the borehole deepened. At the 7.5-mile depth the drill bit began to reach its maximum heat tolerance. Instead of the predicted temperature of about 210 degrees Fahrenheit it was instead 355 degrees Fahrenheit. At that level of heat and pressure the rocks began to act more like putty rather than as a solid. When the drill bit was removed for replacement the hole had a tendency to flow closed.

THE FLOOD EVIDENCE

Nearly all ancient writings describe a great flood, and worldwide scientific evidence exists to support this contention. The most compelling evidence is the worldwide oceanic ridges and trenches that interconnect and circle the Earth (go to the internet and check out a topographical map of the Earth's surface). When water is compressed beneath ten miles of Earth's crust, it would have an unimaginable explosive force if it were suddenly released (relatively easy calculations can show the magnitude and effects of this force). The force would be so great that chunks of Earth would be blasted into outer space (perhaps this is why most of the Moon's craters are on one side). The amount of water released might be enough to cover the entire Earth including its newly formed mountains (perhaps this is why mountains have seashells at their summits). Earth being pushed up at the ten-mile deep trench locations would move land masses apart such that their new locations and newly formed mountains would cause a global mass imbalance that would cause the Earth's spin axis to rotate, thus causing temperate regions to become frozen, and perhaps contributing, along with falling chunks of mud and ice, to mammoths becoming instantaneously frozen in an upright position while eating temperate climate vegetation.

The heat generated from the subterranean water explosion would cause granite in the Earth's crust to become like putty and cause continents to drift apart as a result of water lubrication. When the continents came to a sudden stop due to opposing continental movement, the putty-like granite would form mountains. This phenomenon would occur like the buckling of a train when it crashes into another train coming from the opposite direction. After expelling water from the subterranean void, much of the expelled water would, over a period of time, recede into a reduced-size subterranean void to a new level about three hundred feet lower than where the ocean level is today. At this reduced level, all of the continents would be connected by land bridges and islands would be accessible with primitive boats. This condition would allow the migration of people and animals from one starting location to reach all parts of the globe.

After reaching its lowest point, the oceans would rise up again to their current level, covering the land bridges and separating the continents. This lowering of ocean level and subsequent rise would result from water draining into the subterranean voids and then being pushed back out again as the Earth's crust settled to fill and compress the subterranean void. Drainage from water trapped in large high mountain lakes to the lowered ocean level provides a logical explanation of how, landmarks like the Grand Canyon were formed.

NOAH'S FLOOD

Occasionally the History Channel has a Noah's ark program. To their credit, they do mention the subterranean water theory as one of many possibilities. On the other hand, the theory was followed up with commentary from a geologist who claimed that there is no evidence for subterranean water and that there was no way for this water to form. In other words, the History Channel leaves the impression that the subterranean water theory is bogus. The History Channel's geologist also used subliminal terms like *biblical myth*, *Christian fundamentalists*, and *creationists*, to discredit the worldwide flood and the ark, while seeming to present an unbiased, scientifically based program. I wonder if the geologist spokesperson had ever heard about the Kola Superdeep Borehole!

Because flood "myths" are prevalent in so many cultures, the History Channel program had to admit that some kind of flood must have occurred in antiquity. To provide a scientific explanation, they found scientists who theorized that a localized flood occurred when the Mediterranean Sea overflowed into what is now the Black Sea. The source of the Mediterranean's increased water level was stated to be a result of rising ocean levels when glaciers melted at the end of the ice age. My problem with this theory is that the ice age presumably occurred many thousands of years before the biblical timeframe of the flood. Since glacial melting after the ice age would have predated presumed human existence, how could the flood myth originate with no one to tell the story? Furthermore, if the flood were localized, it assumes that the Black Sea human civilization, with their flood story, spread to all parts of the globe. If this were the case, then global migration would have had to have been by large boats since the continents and islands would have been separated at that time by large bodies of water.

Whether or not God planned to have the flood event happen is a subject of speculation. The Bible implies that God caused the flood to happen because of the evil caused by hybrid humans that existed in the world at that time. My thinking is that God did not intend to release the *"fountains of the great deep"* when Earth was being terraformed. After all, if Adam had not eaten from the tree of good and evil, there would be less reason for the destruction of his offspring; in which case, the subterranean water would ooze up and be filtered through ten miles of the Earth's crust to be released on the Earth's surface as a fresh water mist to water the ground. Genesis Chapter 2 Verse 6 says, *"But a mist used to rise from the Earth and water the whole surface of the ground."* This sounds to me like a built-in sprinkler system and source of fresh water for the idealistic Garden of Eden. No rain required.

Also, with water separating the Earth's crust from its molten core, it's conceivable that volcanoes would not erupt and earthquakes would not occur. With no asteroids or comets resulting from the flood eruption, there would be no chance of their impact with the Earth. In addition, with a uniform global climate and reduced amount of ocean water, the chance of hurricanes, tornadoes, and tsunamis would be less. And finally, but somewhat unrelated to the subterranean water, because Adam was condemned by God to eventually die a human death, the microbic causes of disease and death may not have existed. Perhaps, except for accidental death, mortal human death as we know it was not in God's original plan.

According to the Genesis account, Adam was created from a transformation of dust atoms into a human life form – with a soul. This tells me that God had the ability to communicate with atomic structures and transform them at His command.

Perhaps God thought that humans could have a free will while not being contaminated with evil. If so, mortal human existence could have been much less of a struggle. However, it's interesting to note that all of the angels, *sons of God,* that may have worked on the Earth terraforming project, were also contaminated by the knowledge of good and evil. As stated in Genesis Chapter 3 Verse 22, *"Then the Lord God said, behold the man has become like one of us. Knowing good and evil; lest he stretch out his hand and take also from the tree of life, and eat and live forever."* This could be interpreted as Christ being the tree of life!

TERRAFORMING AND A YOUNG EARTH

Because my reasoned explanation of a global flood parallels what is written in the Bible, it makes sense to me that the Bible's description of events leading up to the flood might also have some validity. These events include recently terraforming the Earth in six days and the creation of Adam and Eve. So in concluding this chapter, I'll examine this possibility.

Although the idea of terraforming the Earth in six days may be a little hard to swallow, even with God's ability to transform matter, many would argue that human artifacts predate Adam and Eve, and dinosaurs and other living species preceded human life by millions of years. My response is: how do you know these artifacts predated Adam and Eve and dinosaurs existed millions of years ago? Of course you would say scientists know this from radiocarbon and radiometric dating techniques. As a counter argument, I would say that these dating techniques are based upon unproven assumptions and do not correlate with observable evidence. And, you would say, what assumptions and what evidence? If you still have an evolutionary mindset, you might also say something less polite.

First of all, the smoking gun to debunk a recently terraformed Earth would indeed be so if human artifacts preceded the existence of Adam, or dinosaurs existed millions of years ago. Scientists, using radiometric dating techniques have shown pottery and other human artifacts do predate Adam, and that artifacts that existed after Adam correlate with each other. If this is true, then human existence may have predated Adam. Additionally, if radiometric dating to determine the ages of dinosaurs is correct, then the recently terraformed Earth scenario is probably incorrect.

The fact is, many observations point to a relatively young terraformed Earth, which may actually be billions of years old, but was recently terraformed to support life. To support the young terraformed Earth assumption, currently used dating techniques may not be valid. For instance, if the rate that radioactive carbon is formed today was different in the past, it would invalidate the radiocarbon dating technique for previously living species. Additionally, radiometric blind testing of artifacts (non-living entities) has shown a wide range of variability. Bias toward finding what the scientist expects to find cannot be ruled out. Also, recent evaluations of the radiometric dating technique have raised questions regarding residual helium and radioactive decay markers called halos.

So what evidence is there for a young, terraformed Earth? Most of the following has been excerpted from Dr. Walt Brown's book, *In the Beginning,*

- The Moon is receding from the Earth at a rate that would place it too close to the Earth if the Earth is 4.6 billion years old. The depth of meteorite dust on the Moon should be much more than what was found. Heat currently being emitted from the Moon indicates a much younger age. Craters on the Moon are relatively recent because of the lack of "creep" which would have, over a long period of time, leveled them.

- Great pressures found in some oil wells would not have been retained for millions of years. Manmade artifacts have been found in coal seams. The absence of meteorites in coal, and other geologically old material, indicate recently made material or reduced meteorite activity in the past!

- Measurements show that the sun is shrinking at a rate of a few feet per hour. If so, it would have been too hot for life to exist on Earth only several million years ago. Also, if the sun was primarily a result of nuclear fusion, the Earth would be bathed in three times as many neutrinos as observed today.

- At the rate that the Earth's magnetic field is currently decaying (a half-life of 1,400-2,000 years), it would have produced an electric current that would have created an excessive amount of heat for life only 20 thousand years ago.

- Meteoric dust is accumulating on Earth so fast that it would be 16 feet thick after four billion years. Also, because this dust is high in nickel the Earth's crust should be high in nickel – which it is not.

- Evolutionary population growth models predict human and animal populations that are much greater than exists today -- see my previous commentary.

- Radiometric dating that relies on nuclear decay, is contradicted by test samples showing high amounts of helium -- generated from nuclear decay. For example, uranium based nuclear decay samples have indicated an age of 1.5 billion years. However, the amount of residual helium indicates an age of 6,000 +/- 2,000 years.

- When a radioactive atom decays it gives off energy at a characteristic level. The energy burst damages the mineral matrix, which leaves behind a circle called "pleochroic halos." By observing the halo array one can deduce the make-up of the parent material when the mineral was formed. When radon-222 decays into polonium-218 it takes a half-life of only 3.82 days. Further degradation steps, leading to lead, are also very rapid. Amazingly, the set of halos characteristic of polonium isotopes is sometimes found without the more slowly forming uranium isotope halos, showing no evidence of a parent cluster of uranium – just polonium. Apparently there never was a uranium cluster present, and the original cluster must have been polonium. A result that indicates a recent age for inorganic materials.

• Scientists claim that the formation of fossil fuels like coal, oil, and natural gas, is the result of 100's of millions of years of decaying animals and plant life. However, if this were true there would be a gradual build-up that would be uniformly located around the world, and not the randomized and buried clumped formations that have actually occurred. One theory for randomized, buried, and clumped fossil fuel formations stems from a belief that during a worldwide cataclysmic flood event, trillions of plants, trees, and animals were washed into clumps, suddenly buried, and sandwiched between sedimentary rocks. Recent experiments suggest that the fossil fuel formations do not require the claimed 100s of millions of years. In fact, one U.S. Bureau of Mines experiment showed that oil can be produced from organic material in only 20 minutes.

- Currently, cosmic radiation strikes the upper atmosphere and converts about 21 pounds of nitrogen into radioactive carbon (carbon 14) each year. Most of the carbon 14 combines with atmospheric oxygen to form radioactive carbon dioxide. Plants absorb the radioactive carbon dioxide and living animals eat the plants. When the living animal or plant dies its radiocarbon decay is no longer balanced by intake. Since the half-life of radiocarbon is 5,730 years we can estimate the date that the living animal or plant died. If the carbon 14 to carbon 12 ratio were different in the past, the dating technique would be invalid. With a larger land mass, and a greater amount of vegetation before the flood, this ratio would have been dramatically different, thus creating the illusion of a much longer age. Moreover, since carbon 14 is found in all fossilized animals -- including dinosaurs -- it would be impossible for these animals to have existed more than 100 thousand years ago since our measuring methods could not detect the infinitesimal amount of remaining carbon.

The above information is only the tip of a large scientifically based and remarkable iceberg of data that supports the creationist point of view. If you are scientifically inclined, I think that you will be amazed at what the creationists are saying about the fallacy of current dating techniques and the formation of comets and asteroids. To give the reader some idea of what I mean about comets and asteroids -- the following information was excerpted from Dr. Brown's book:

There is a NASA photo showing an asteroid named Ida and its moon, Dactyl (re: my book "Reaching America's Destiny"). In 1993, the Galileo spacecraft, heading toward Jupiter, took this picture 2,000 miles from asteroid Ida. To the surprise of most, Ida had a moon (about 1 mile in diameter) orbiting 60 miles away! Both Ida and Dactyl are composed of earth-like rock. We now know that at least 230 other asteroids have moons; nine asteroids have two moons. According to the laws of orbital mechanics, capturing a moon in space is unbelievably difficult—unless both the asteroid and a nearby potential moon had very similar speeds and directions and unless gases surrounded the asteroid during capture. If so, the asteroid, its moon, and gas molecules were probably coming from the same place and were launched about the same time. Within a million years, passing bodies would have stripped the moons away, so these asteroid-moon captures must have been recent.

From a distance, large asteroids look like big rocks. However, many show, by their low density, that they contain either much empty space or something light, such as water ice. Also, the best close-up pictures of an asteroid show millions of smaller rocks on its surface. Therefore, asteroids are *flying rock piles* held together by gravity. Ida, about 35 miles long, does not have enough gravity to squeeze itself into a spherical shape.

SUMMARIZING

The fountains of the great deep could have launched rocks into space as well as muddy water. As the rocks moved farther from Earth, Earth's gravity would become less significant to them, and the gravity of nearby rocks would become increasingly significant. Consequently, many rocks, assisted by their mutual gravity and surrounding clouds of water vapor, would merge to become asteroids. Isolated rocks in space are meteoroids. Drag forces caused by water vapor and thrust forces produced by the radiometer effect would concentrate asteroids in what is now the asteroid belt. All the so-called "mavericks of the solar system" (asteroids, meteoroids, and comets) could have resulted from the explosive events at the beginning of the flood.

The above information, and much more, can be found in Dr. Brown's book (The 8th edition is free on the internet, so there is no excuse for not checking it out). The scientific evidence for a young Earth is compelling, and is confounding and confusing to mainstream scientists who can't seem to escape the confines of the box in which they have become trapped. I wonder what they will say if a pre-flood manmade artifact were to be found on the Moon or on one of the asteroids! Since the scientific basis supporting a recently terraformed Earth flies in the face of current scientific doctrine it will take time, possibly generations, for it to take hold.

I believe future exploration and possible findings on the Moon and asteroids will begin the process of re-establishing God as the creator of all things. In my opinion, something will be discovered that will change much of what current mainstream scientists, and their indoctrinated followers, think they know. When people are faced with irrefutable evidence of God's existence, the tide of public and scientific opinion may begin to turn. This is why I believe Christ's return is not as imminent as many people think. We need time for new scientific discoveries to evolve and become as much a part of future thinking as Darwin's theory of evolution is today. By understanding the Biblical flood, and all of its ramifications, one might conclude that the Genesis account of creation is not as much a fairy tale as most people have come to believe.

In concluding this commentary, I found one quote to be quite interesting. One of the reviewers of Dr. Brown's book, a senior electro-mechanics research scientist at the University of Texas at Austin, stated, *"If I were to send my child off with only two books, it would be the Bible and In the Beginning."* In my opinion, Dr. Brown's book may be the most important scientific book ever written! It is the most complete reference work I have ever encountered on the scientific aspects (including substantiating mathematics) of our origins, and it makes sense out of many unrelated and contradictory facts about the universe, our solar system, and planet Earth.

CHAPTER 8

IS THERE AN AFTERLIFE?

The data is overwhelming that a creative spiritual force does exist in the universe, and for convenience let's call this spiritual force God. If God exists, then logic would dictate that humans must be a special creation with a special purpose. Since God provided humans with a free will, they can either accept or reject God. If God created humans without a free will, what would be the sense of it? They would just be obedient humanoid robots who could never appreciate or enjoy God's creations. On the other hand, I believe God probably intended for humans to have a free will that was devoid of evil.

Imagine if you were a great artist with no one to appreciate or enjoy your work of art! The risk is that with an audience that has a free will, your work of art may not be appreciated or enjoyed. In fact, it may be criticized and/or ignored. This is the risk God has apparently taken. If God exists, and we are God's special creation, it makes sense that God would provide instructions. I have come to believe the Bible was divinely inspired to help us understand God's eternal plan and to provide guidance on how we should live our lives. After all, God's Ten Commandments, from what I can tell, were not meant to be ten suggestions.

Now you may ask, are humans intended to just live for a short length of time and then disappear, or is it more rational to think God would want humans to live an eternal life? Forgetting for a moment that the latter option is explicitly endorsed by the Bible, why does it make sense for humans to live eternally? Being an engineer, I've developed a need for things to be logical and rational. While my knowledge of God's intent is meager at best, I believe God has provided clues that verify what the Bible states. First, we are without a doubt special creations of God, and in a sense we must be God's children. I say this because we possess the ability to communicate with God. If you are a parent, you instinctively love your children and want them to have more than you might have had in your life. Where did this parental instinct come from, and why wouldn't God treat His children in the same way?

THE BIBLE

If we are going to argue in favor of an afterlife, the Bible is a good place to start. While some biblical scholars claim to know what the afterlife is, I want to know more, especially about what occupies an eternity of time.

Let's begin with. Mark chapter 12 verse 25, which states, *"For when they shall rise from the dead, they neither marry, nor are given in marriage; but are as the angels which are in heaven."* Revelation chapter 21 verse 4 states, *"And God shall wipe away all tears from their eyes; and there shall be no more death, neither sorrow, nor crying, neither shall there be any more pain: for the former things have passed away."* In Billy Graham's book *Death and the Life After*, he interprets 1 Corinthians chapter 15 to say, *"Yes, they are weak, dying bodies now, but when we live again they will be full of strength. They are just*

*human bodies at death, but when they come back to life they will
have superhuman bodies."*

THE SOUL

If we're going to continue exploring the possibility of an
afterlife we need to provide a scientific and rational basis for
the human soul. Since, if an afterlife exists, it makes sense
to assume that the soul continues to live on after the mortal
body dies.

The Computer Brain

I'll begin with the assumption that the soul and the brain
are two separate entities, even though the current mainstream
scientific community doesn't see it that way. Of course their
contention is based upon evolution and has nothing to do with
God. If you do not believe in God, then the soul, or mind,
must have evolved and is fully integrated with the brain. After
all, if we cannot see the soul, then it must not exist. It's that
simple. But is it?

If you look at the brain as being a complex biological
computer containing over 30 billion cells, then certain aspects
of human or animal instincts and functioning can be included
in the programming. On the other hand, the human aspects of
awareness, original thoughts and emotion, are much different.
How can a computer be aware of itself, or have the emotions
of achievement, appreciation for music and beauty, being in
love, grief for the loss of a loved one, or hope? The high-speed
calculating capabilities of a computer have little meaning if
not guided by an inquisitor. For all intents and purposes, the
brain is a just a physical housing for interconnected neurons
that perform according the way they are programmed. As the

computer/brain matures and alters its behavior because of encounters with the physical world, it is, in effect, adjusting to its environment and memories of past events, a form of artificial intelligence or self-learning. It's not beginning a process of developing the feelings of achievement, love, and hope. It's my opinion that no self-learning computer, regardless of how sophisticated, can ever attain this capability. Although computer-based memories of encounters with the physical world, and learned subjects, are the data base for thinking and reasoning, they don't formulate the questions.

Consider this. If the mind is reduced to the physical brain, feelings become chemical reactions, attractive objects become light waves, and beautiful music is nothing more than vibrating molecules. How can a computer feel what it is analyzing? Is being in love just a chemical reaction, or is it something else? Is it the non-physical soul that is causing this emotion? And what about decisions you make? Are your decisions predetermined by a computer program, or do you have a free will to make a choice? If you have the free will to make a choice, then you can override the computer's decision. If your decisions are controlled by computer analyzed physical events outside of your free will, then you do not have a free will at all. What is the process that allows you to override your brain/computer? Is it a soul?

What if the brain/computer is damaged and the soul is unable to evolve normally? In this instance, it appears that the soul needs to work in conjunction with the brain's computing and bodily control capabilities in order to evolve and mature. It's not a matter of having an inferior soul. In fact, if God has provided each human with an immortal soul, my thinking is that all immortal souls are created equal, but physical brains are not. So, when the soul leaves the brain upon physical death, it may not be sufficiently mature, and my guess is that God needs to compensate in some way. And, if the soul is immortal, what do you think that is?

Androids

To make my point about the separateness of the soul and the brain, let's examine *Star Trek's* android, Data. As you may know from watching *Star Trek: The Next Generation*, the android named Data has superhuman strength and intelligence, but lacks the human attribute of emotion. Data is an advanced computer that can interact with humans and perform extraordinary tasks, but is constantly confused by human emotions such as fear, love, anger, jealousy, and grief.

From a technical standpoint. Data would have the senses of sight, hearing, smell, and touch. The sense of taste would be closely linked to smell, in that he might have the equivalent of a mass spectrometer that can define elements and through computer interpretation discern the nature of a substance. The capability of seeing might include three-dimensional video camera-type observations of his surroundings that are interpreted as graphically rendered objects. Seeing could also be enhanced to include X-ray vision. With further refinement, the material nature of the objects could be identified. Hearing would be in the form of vibrating air molecules, and with further refinement the directionality of the sounds could be discerned. Data's body could also include miniature mechanical nano-sensors that would provide the sensation of touch.

Occasionally Data plays poker with some of the crew members. At one point, Data throws his cards in when one of the crew members bluffs Data into thinking that the crew member had a winning hand. This confuses Data, since the bluff appears to be irrational. When the bluff is explained to Data, he is able to incorporate the play into future poker games. I find this interesting since I have been fascinated by playing Scrabble on my computer. The computer not only has instant access to the entire dictionary, but it also has the capability of employing strategy. While playing the game, I have the feeling that I am playing with another human. However, because the computer has a distinct advantage, I need to select a level of play that suits my competence level. In other words, the computer scales back its capabilities to allow me to compete. My question is not whether the computer can beat me at the game, but rather, is the computer capable of inventing a game that humans would enjoy? Furthermore, why would it invent a game in the first place? I can't imagine that Data derives enjoyment from playing poker, since he is incapable of the feeling of enjoyment. And, without understanding the feeling of enjoyment, how could he design a game with human enjoyment as the objective. So while watching Data in his role on Star Trek, consider that he is an extraordinary machine, and not a human. There is no amount of computer programming that will produce original thoughts or human emotions. Could it be that original thoughts and human emotions come from the soul?

In one Star trek episode Data discovers that his inventor had made another android named Lore, but Lore has an emotion chip. Having human emotions, superhuman strength, and a superhuman brain, Lore decides to eliminate the inferior human species and create others like himself. This is a problem that some fear as android invention evolves. However, this fear is ill founded, since it is inconceivable that any inventor could ever produce a so called emotion microchip. And, without human emotions that contain evil thoughts, androids would not have a desire to overthrow humans. The android could be programmed to act based upon a decision tree that endlessly mimics human behavior, but it would not be acting based upon human type emotional feelings. This is an important concept, and one worth remembering; and here's why.

If you think that androids can be developed with an ability to have emotions and a free will, you would need to mathematically define esoteric terms like love, hate, ambition, self-awareness, jealousy, greed, grief, happiness, joy, sadness, curiosity, embarrassment, desire, appreciation, compassion, altruism, fear, enjoyment, intuition, and the emotions of crying, blushing, sulking, and laughing. You would also need to define the process of original thought and reasoned logic. Because these terms cannot be defined mathematically, they cannot be programmed. If they cannot be programmed, then what are they? Where do they come from? How can they evolve, if there is no predecessor? If all living species developed over a similar evolutionary timeframe, then why are humans unique? Did the emotion of love develop before the emotion of hate, or did all of the human emotions develop simultaneously?

An evolutionist would say that it's just a matter of time when the dog will acquire the missing link in its brain. Really! If so, then we are not just talking about dogs; we are also talking about insects. Will insects become humanlike in time? I don't think so, but this is what you are being told by scientists who believe in evolution of the brain without the influence of God.

Pavlov's Dog

Let's look at the soul and brain another way. As previously mentioned, I believe that the soul is unique to humans, and that animals do not have a soul. If animals don't have a soul, then how does their brain function? To begin let's assume that animals have instincts and bodily control functions programmed into their biological computer/brains. In some cases, animal instincts are so amazing that they are beyond human comprehension. Fish returning to their spawning location and birds flying south for the winter are mysterious examples of what I mean. Beyond instincts, is the ability to learn and react. Pavlov's dog is experimental proof that some animals can learn to perform in a certain way based upon stimuli. If a dog does something that the owner wants it to do, the dog gets rewarded with food and caring. With enough trials, the dog will repeat the desired action. The dog also connects its owner with food and caring. When the owner goes away the dog appears to grieve. When the owner returns the dog appears to be happy. Does this mean that the dog loves its owner, or is the dog just reacting to stimuli? To put this in human terms, how many times have you seen a male dog fall in-love with a female dog? Does the dog appreciate beautiful music? Does it grieve when its mother dies? Does it prefer one color over another? Has it ever created anything? Does it show its owner that it can do new and better tricks than it had been taught? Does it worry about other dogs in its family? Does it exchange ideas with other dogs? Does it teach other dogs what it has learned? Does it indicate that it would like to drive a car? Does it have a desire to win a ball game, or does it just want to play ball? Have you ever seen a dog that was embarrassed? Have you ever seen a dog that was jealous? I don't know if this makes my point, but I think that you will agree, that humans obviously have something that animals don't have. Is it a soul, or is it something that dogs will acquire if given enough time to evolve?

Soul Science

Scientists are currently examining animal and human brains to determine what makes them tick. They see living cross sectional color renditions of human brain activity, and correlate this activity with emotions and bodily functions. They dissect deceased brains in an attempt to ascertain why some brains behave differently than others. They run tests to evaluate how brains respond to various stimuli. To me this is analogous to trying to examine a computer's microchip to determine the nature of the program that it's running. While we may be learning more about how a biological computer works, we are probably never going to understand the emotional function. While we know empirically that the emotion aspect is there, I don't think we will ever find it.

However, in spite of what I've just said, there's a book entitled *The Brain That Changes Itself* by psychiatrist Norman Dodge. This book tells how leading doctors are using cognitive treatments to help children overcome learning disabilities, senior citizens to improve memories, and paralyzed stroke victims to speak again. These treatments indicate that the soul/mind is capable of rewiring the physical brain. Thus illustrating the separateness of the two entities. If your soul/mind is capable of changing the brain, it's reasonable to assume that the soul/mind could survive the death of the brain.

One of the most telling of scientific experiments has been to measure weight differences before and after human death. In one instance six patients in a nursing home were wheeled onto a scale just before their death. The scale measured to within one gram, and the results were startling. In each case the weight before death and after death was reduced by an average of 29 grams (454 grams equals one pound). Similar tests have been conducted worldwide with similar results. If you are a scientist, you would have to conclude that something is leaving the body at the time of death. Could this be the soul? Could this be the undetectable invisible ingredient that creates human emotions?

AFTERLIFE EVIDENCE

Let's now look at the scientific evidences that support the existence of an afterlife.

Near Death Experiences

Oncologist Dr. Jeffrey Long has written a book entitled *Evidence of the Afterlife*. For over a decade, Dr. Long cataloged stories from over 1,600 people who have had the near death experience, and concludes that an afterlife must be real. Patients whose heartbeat and breathing had stopped and their brain had ceased to function, report remarkably consistent stories that are generally lucid and highly organized. Dr. Long states that, *"if these people had no brain function, like you have in a cardiac arrest, I think that it is the best model we're going to have to study whether or not conscious experience can occur apart from the physical brain. The research shows that the overwhelming answer is yes."*

I find Dr. Long's statement about the brain not functioning to be quite fascinating. I find it fascinating because the skeptics of life after death always attribute the dying conscious or subconscious thoughts to a residual functioning of the brain. But, if there is no brain function, how can this be? My problem with the *"out of body experience"* is not that the soul leaves the body at death, but it's the claim that the out of body entity is able to see and hear what is going on. Perhaps the soul has a form of sight and hearing that doesn't require eyes and ears as we know them. In any event, Dr. Long's study does support what I have been saying to this point, that the soul is a separate and distinct entity that survives after physical death.

Extra-Sensory Perception

Although I'm reluctant to include extra-sensory perception testing and evaluation in my discussion to show evidence of an afterlife, I would be remiss if I didn't at least mention it. My reluctance comes from my own skepticism, since the physical means by which thoughts are transmitted is not measurable by any known scientific means. But, like trying to find the soul, just because it can't be measured, doesn't mean that it doesn't exist. In fact, despite many shortcomings in scientific rigor related to currently available paranormal data, there remains an element of uncertainty that cannot be dismissed at this point in time. I find it interesting that even though the vast majority of scientists do not believe in the paranormal phenomena, because they believe that the data is biased toward the investigators premise, that they themselves are guilty of the same bias when it comes to their premises.

Generally speaking, the categories of worldwide extra-sensory perception testing are:

1 *Precognition* – knowledge a person may have of another person's future thoughts or of future events.

2 *Psychokinesis* – a person's ability to influence a physical object or event, by merely thinking about it.

3 *Telepathy* – a person's awareness of another's thoughts, without any communication through normal sensory channels.

4 *Clairvoyance* – knowledge acquired of an object or event without the use of the senses.

My opinion is that precognition is not possible because it assumes that future events have already occurred and that the free will of people cannot influence it. Well known psychic Edgar Cayce claimed that some of his predictions did not happen because people's free will prevented them from happening. I find it difficult to accept this argument since it assumes that all that will happen, has already happened. If this were so, then our existence and the existence of the universe is like a game that has already been played, but the results keep changing based upon variations of the free wills of the players on the game board.

On the other hand, I do believe in Biblical prophecy based upon the accuracy of events that were predicted hundreds of years earlier. In this case God is forecasting future events not humans. How God does this is certainly a mystery, especially when you consider humans having a free -- and unpredictable – will, where almost any outcome is possible. Because of this, one might conclude that everything that's going to happen – has already happened. On the other hand, God might be capable of influencing the free will of people and circumstances that fulfill His prophesies. As a result, He proves His omnipotent power and the authenticity of the Bible. Hmm – interesting!

I also don't believe in Psychokinesis, since measurable forces would be required to move objects. On the other hand, if matter had intelligence that could be influenced by the use of access codes, we are into a whole new ballgame. Perhaps human thought is a force that we currently cannot measure.

Telepathy and clairvoyance are not so difficult to dismiss. My own personal experience involved a magician who asked me to think of a playing card. When I thought of the five of diamonds, the magician picked that card out of the deck and asked if that was the card. I was astonished by this magic trick. How could the magician know what I was thinking? Perhaps there is a good explanation, if I knew how to do the trick, but it's not like picking the card from the deck and the magician guessing the card. This somehow involved my mind.

My thinking on the subject of telepathy and clairvoyance is that current researchers are probably "boxed in" by their effort to find a means of thought transmission. They assume that thought transmission is like radio waves that are transmitted by a sending and receiving source, but what if everything in the universe is somehow interconnected and thoughts are intertwined in a universal web of knowledge. Before you say that this is impossible, consider how radio waves and submicroscopic particles are currently being transmitted through the air and through your body. If this were not so then your radio, television, and cell phone would not work. And, when it comes to things that scientists do not understand, consider the previously mentioned dark energy that is causing the universe to expand and accelerate.

From what I've said so far you know that I believe there is a distinction between the soul (or the mind) and the brain. And, that's why I've spent so much time in this chapter to discuss it. If there is no soul, then there is nothing. A human lives, eats, sleeps, works, plays, procreates, and dies. As Shakespeare stated, *"a poor player that struts and frets an hour upon the stage and is heard no more.* "On the other hand, if there is a soul, then the idea of eternal life and god-like capabilities still remain on the table. Just because you cannot see the soul, doesn't mean that it isn't there. But in your heart of hearts you must know that there is something there, and it is you. It is the essence of who you are.

THE PHYSICS OF IMMORTALITY

In a book authored by Dinesh D'Souza entitled *Life After Death – The Evidence* I was fascinated by his chapter on the physics of immortality. In this chapter D'Souza tells about recent findings by mainstream scientists that are inadvertently and reluctantly discovering God as the only explanation for what is. In an apparent panicked attempt to disprove God, the previously mentioned multiple universes theory has recently taken center stage. An infinite number of universes helps to explain the statistical impossibility of what exists. As stated by D'Souza, *"the probability for what exists is like buying a lottery ticket in each of America's 50 states and winning them all."* Even if the unprovable multiple universe theory is correct – the explanation for who or what created them would still be a mystery. Again – if there is a creator God, an afterlife is not only possible – it's probable.

So what are these recent scientific findings? Well consider this for openers. It has been shown that a particle in one location can affect the behavior of a particle in another very distant location with no apparent means of communication. Other experiments have shown that subatomic particles in atomic structures behave differently according to what each observer thinks – an apparent form of the above mentioned psychokinesis.

Speaking of atomic structures have you ever thought of matter (for instance a table) as empty space. The major part of an atomic structure's mass is in the nucleus, but it's size – relative to the size of the atom – is like the size of a baseball compared to the size of Fenway Park. Most of what comprises the atom is empty space that is comprised of a few infinitely small electrons, quarks, and other subatomic particles!

If atomic structures weren't strange enough, how about asking what would the speed of light be if you were traveling at the speed of light and measured it? The answer is that it would still be the same as if you measured it while standing still. This is counter intuitive when you consider traveling at 50 miles per hour on a train and measured the speed of a passing train going 60 miles per hour. Because of relative speeds you would say that the passing train was going 10 miles per hour.

Wait! I'm not finished. Aside from the previously discussed dark energy and matter that's causing the universe to not only expand but is accelerating the expansion, scientists now tell us that there might be more than the four dimensions of length, height, width, and time. M theory suggests that there are ten dimensions of space and one of time. If so, existence in a life after death might include these hidden dimensions. In any event, to assume that what we see is what we get – is like believing that the earth is flat. It's old science.

Finally, with regard to statistical impossibilities, Steven Hawking's book, *A Brief History of Time,* mentions that the nature of the universe is remarkably sensitive to six numbers. If you imagine setting up a universe by precisely tuning six dials to yield a universe that could harbor life you might start to get a feel for how impossible this might be. It's called the Anthropic Principle and is accepted by today's leading scientists.

So if you believe in one universe, there is no logical or scientific reason for not believing in God. And, if you believe in God there is a very good possibility that life goes on after death. If you believe in multiple universes, there is still no logical or scientific reason for not believing in God!

THE AFTERLIFE AND CHRIST

Finally – perhaps the best evidence for life after death is the resurrection of Jesus the Christ. Do your own research regarding the accurately prophesized life of Christ (more than four hundred years before Christ in the book of Daniel) and His historical miraculous life, death, and resurrection (including more than five hundred witnesses) for more evidence of an afterlife. However, be forewarned – what you find might change your life.

CHAPTER 9

AMERICAN EXCEPTIONALISM

've included the American exceptionalism topic in this last chapter because I want to convey a very important message to those who believe America is not exceptional and that it needs to be condemned and punished for its past wrongdoings. Rather than use a term *exceptional*, which can be misinterpreted to mean arrogant people who took advantage of people in other countries and are just lucky to be where they are I would instead use the word *destiny*. The word *destiny* implies a creator God who has made America what it has become and what it is intended to become. In relation to what the world was before America came into existence, one could argue that America has indeed been exceptional. And, I would argue that it will continue to be exceptional until it fulfills its destiny. Let me explain.

AMERICA'S PURPOSE

Does America serve a purpose in God's cosmic plan? I believe it does, and that a series of miracles have allowed the United States of America to come into existence. After studying America's history, I believe that is exactly what has and is happening. The miracle that impresses me most is the life of George Washington; and of the many miracles surrounding Washington's life, the one I'm most impressed with is when he was shot at point blank range by Native Americans early in his career but still survived. The reason I'm so impressed by this miracle is because of a retelling of the story by the Indian chief who led the attack and who traveled a long distance to visit Washington fifteen years after the event. The chief said that during the battle he commanded his warriors to shoot the officer in charge. After many shots were fired, the chief noticed, that the officer could not be killed. It was at this point that he told his warriors to stop firing at him, because he believed that the *Great Spirit* was protecting the officer for some great purpose. In spite of four bullet holes in his coat and having two horses shot beneath him, Washington had no wounds on his body. The chief further recounted that one of his warriors said, *"I fired seventeen clear shots and couldn't bring the officer to the ground."*

I recommend that you read about Washington's life to fully understand how the *Great Spirit/God* had protected him and allowed America to come into existence. You will learn that British General Howe had assembled thirty-two thousand troops against Washington's eight thousand, yet Washington and his troops escaped because of a miraculous fog. You will learn that at the battle of Yorktown, British General Cornwallis was about to receive reinforcements, but a violent storm caused the British ships to be blown down river and divide his forces. Later that day, Cornwallis surrendered to Washington. And the list goes on from events that happened at Valley Forge to his crossing the Delaware River. The miracles surrounding George Washington's life, including a *mysterious visitor* that he encountered while encamped at Valley Forge, are a strong indication of divine intervention, but there's much more.

The evidence of God's interest in America is overwhelming. It begins with Christopher Columbus and the Pilgrims to saving Ronald Regan's life to go on and win the Cold War. Pivotal battles at Gettysburg, New Orleans, Midway, and events that occurred during the Lewis and Clark expedition changed the course of history. The facts surrounding the miraculous beachhead landing at Normandy, which prevented Hitler from ruling the world, and the miracle surrounding the writing of our Constitution bears testimony of a divine being acting on America's behalf.

I don't think anyone would deny the miraculous technological progress that has been made during America's existence, or the crucial roll it has played in keeping the world from being ruled by tyrants like Adolf Hitler. From what I have experienced and observed, I believe America has a destiny that has yet to be fulfilled. How about America leading the world to a better place? How about a place where all men (and women) are created equal and that they are endowed by their creator with certain unalienable rights, that among these are life, liberty, and the pursuit of happiness? How about a place where all of the Earth's people have freedom from tyranny and access to modern-day conveniences and medical achievements?

Although America has experienced many growing pains, it still stands as the greatest hope for mankind. As President Reagan once said, *"We are a shining city on a hill."* In my view, we are, and will continue to be, a beacon of light for the rest of the world to follow.

As I see it, our growing pains were meant to teach us lessons that have helped us to mature as a nation. In the past we learned that slavery and discrimination were wrong, and women should vote and have equal rights. We learned that the cost of freedom comes at a high price, and we're now in the process of learning, the hard way the importance of free enterprise and how government can't spend its way to prosperity.

As of this writing, the United States is heading for bankruptcy and hyperinflation (the national debt is now approaching $20 trillion). If this happens, it will be the end of America, and its destiny will not be fulfilled. I believe God will not let this happen. From my perspective, no other country in the world can save mankind from an inevitable fate of tyranny, poverty, hunger, and disease. The United States has the type of government, diversity of population, natural resources, climate, land, and historical maturity needed to save the world. And, I believe God has ordained it to be the catalyst that will help fulfill His plan for humanity.

So what is America's destiny? Where do we need to go that is so much better than where we are now? Well, what about helping an exponentially growing world population of poverty-stricken people to enjoy modern conveniences and medical advances? How about curing cancer and other life-threatening diseases? How about setting an example that will help free people from tyrannical dictators and human bondage? And, if we are talking about destiny, how about spreading the word of God? After all, we are a Christian nation that was founded on Christian principles. With this thought in mind, I recommend that you read two books authored by Peter Marshall Jr. entitled *The Light and the Glory*, and *From Sea to Shining Sea*, to understand what I mean. These books tell of America's miraculous history beginning with Christopher Columbus, and the Pilgrims to our westward expansion to the shores of the Pacific. However, there's much more that you need to know.

FREE ENTERPRISE

Not too long ago I watched a History Channel presentation entitled, *The Men Who Built America*. Men like Vanderbilt, Morgan, Rockefeller, and Carnegie changed the world and became extremely wealthy in the process. They amassed fortunes so huge that in today's dollars would make Bill Gates look like a second-rate billionaire. They were ruthless in their quest to make money, but as a result they literally built America with their investments. Their monopolies in railroads, steel, and oil stifled competition by crushing anyone who would oppose them. When Teddy Roosevelt became president, he used the power of the government to break up their monopolies. For instance, Rockefeller's Standard Oil Company was broken up into many smaller oil companies which today are known as Exxon, Mobil, Shell, and others. Ironically, Rockefeller knew that with the advent of Henry Ford's automobile that future use of oil, beyond kerosene lanterns and heating, was destined to earn a great deal of money, so he invested his money in the new upstart oil companies and eventually became even more wealthy! Regardless of what contempt, disdain, and jealousy some people might harbor for these men, they all became prolific philanthropists in their later years, and their contribution to the growth and prosperity of America can't be measured.

When the government acted to break up the monopolies, it opened the door to new era of capitalism and free markets and the phenomenal growth that America has experienced. In the meantime, under the guise of helping the poor and disenfranchised, the power of the government has also grown to where it is controlling most of their lives. The land of freedom and opportunity is quickly becoming the land of dependence. *We the People* are no longer in charge of our destiny, since many of the dependent people vote for politicians who will provide them with additional benefits. This is a system that will inevitably fail. And, if it does fail, the dependent voters will understand that it was all just a lot of smoke and mirrors. Don't get me wrong; I believe that government should provide a safety net for those who truly need it, but it also needs to encourage and assist those who are able to help themselves.

By empowering the people, we can restore the American dream, and help able-bodied people to achieve a fulfilled life based upon opportunity, freedom, and prosperity and not dependence. When the men who built America risked their own money to create wealth, they also provided millions of jobs and economic growth. When our government began to grow, it had no interest in operating efficiency and using its money wisely. That is why free enterprise companies like Federal Express and UPS can run circles around the government-run US Postal Service. When the government pays $500 for a hammer (which it has), you know that the bureaucrat making that purchase didn't care about how taxpayer money was spent.

SAVING THE WORLD

To the hopeful observer, the opening and closing ceremonies of the Olympics represent how the world should be: all races, cultures, and countries coexisting and living in peace and prosperity. Unfortunately, the reality is much different. Most of the world lives with poverty, hunger, disease, and violence. To make matters worse, many of these people have access to the internet and Hollywood movies, which makes their plight even more exasperating because they can see what they don't and can't have. As the old saying goes, *"what you don't know can't hurt you."*

I believe that part of America's destiny is to rescue the world's poor, but it will require a lot more than we are doing now. Currently, the United States provides foreign aid to many countries, yet the result is increased poverty. In fact, with the populations of these poor countries increasing at an alarming rate, the situation is getting worse since more and more farmland is becoming unproductive and fresh water is becoming progressively scarce. To make matters worse, if foreign aid money is actually spent to help the poor, it's mostly in the form of food that, once consumed, leads to a need for more food. It's obvious that the fundamental problem is not being addressed.

If America is really going to help the poor, it needs to address the underlying causes of poverty rather than treating its symptoms. When the underlying causes are addressed, the problems of overpopulation and threats to world peace will be greatly reduced. But herein lies a dilemma, because treating the underlying causes requires a vast increase in use of the world's non-renewable resources. In other words, if we allow the poor to become legitimate members of the modern-day technological community, it means giving them full access to livable housing, energy, fresh water, material goods, and medicine. It also means giving them freedom, education, jobs that pay a living wage, and access to investment capital. While this may sound like an insurmountable task, I believe that there is a way to make it happen. It's a gradual process that starts with America!

In order for America to rescue the world's destitute people it must first rescue itself. The United States is currently heading for bankruptcy and hyperinflation, and something drastic needs to be done now to reverse this trend. Surprisingly, in my opinion, the solution requires help from the world's destitute people! While you may be saying this sounds absurd, hold on while I explain.

ECONOMICS 101

Years ago famed economist Milton Friedman visited China and observed hundreds of workers building a road with shovels and pickaxes (an example I've seen firsthand during my tour of duty in Vietnam). When Friedman asked the man in charge why they weren't using modern-day equipment, the supervisor said that by using shovels more workers could be employed. Friedman responded by asking, *"why not use spoons and employ even more workers?"* If you believe that spoons or shovels are a good way to keep people employed and eliminate poverty, think again.

If one hundred workers earned $5 per day with shovels, then five workers could earn $100 per day if they used modern road-building equipment. Moreover, these more productive workers could probably build the road ten times faster. This means that the cost to build the road would be much less than the original budget, and these five more productive workers could be paid as much as $300 per day or about $75,000 per year.

But what about the other ninety-five workers, won't they be unemployed? Well, the five productive workers could employ them by having them build five houses and provide other goods and services since these highly paid productive workers now have enough money to pay back an amortized loan, which leverages their money. If the house-builders use modern tools and equipment, they can also earn a good wage, since the faster they build the house the more money they can make. Since they will earn more money than they did using shovels to build the road, they can buy more goods and services that require more people to be employed. Then these newly employed people also buy houses, goods, and services that eventually allow the ninety-five workers to continue working – but with much higher wages.

So here's the point I want to make. It is in the best economic interest of America to help the world's poor. Why? Because the world's poor represent a new consumer base that will grow its economy. Years ago, I remember watching a televised interview with Jack Welch, the former CEO of General Electric, and arguably the greatest private business CEO of all time. When Jack was asked if China was an economic threat to the United States, he said no, since from his viewpoint China represented more than a billion new consumers. As China's economy grows, so will the number of consumers who can buy goods made in the United States. In other words, economic growth in the underdeveloped world will benefit all of us.

With the advent of robotics and worldwide recycling (explained in my book *Reaching America's Destiny*), low-cost manufacturing could return to the United States. When it does, the United States will prosper economically and have an economic incentive (increasing its consumer base) to help the world's poverty-stricken people in a revised form of foreign aid. I say revised because of this anonymous quote, *"America's current foreign aid program is the process of poor people from rich countries giving money to rich people in poor countries."* Let's turn this around by providing equipment and resources that actually help the poor people in poor countries to prosper.

In order for America to grow and prosper, it needs more consumers who will buy American-made goods, and destitute people represent a new consumer base. America can also grow and prosper by transitioning to hydrogen as a portable energy source, and fusion and/or breeder reactor fission nuclear for stationary energy (this process is also explained in an earlier chapter in this book and my book *Reaching America's Destiny*).

CREATING NEW CONSUMERS

Since most destitute people live in tropical, or semi-tropical locations, solar energy can be used quite effectively. Let's turn their overused, nonproductive farmland into productive energy-producing land. Let's then use most of this energy to generate hydrogen from water. The hydrogen can then be sold to the world market and the destitute people can make a profit. It's as simple as that, but it's only a beginning.

For food, let's grow crops hydroponically so less land and water will be required; soil will not be degraded from overuse; weather and flood damage will be eliminated; and pesticides will not be needed.

Now let's provide productive jobs. How about recycling jobs whereby almost all of the materials used in obsolete manufactured products are preserved? This sounds a lot better than just throwing these obsolete products into landfills and making their replacement more expensive and scarce. Read my book *Reaching America's Destiny* if you are interested in how this can be done.

AMERICA'S DESTINY

Ultimately, America's destiny can be fulfilled by engaging in an aggressive space program which includes mining the sky. And, if you read the prologue to this book, you know how that might turn out.

I'll end this commentary with another quote from George Bernard Shaw,

"The reasonable man adapts himself to the World: the unreasonable one persists in trying to adapt the World to himself." If you are a student of history, you might conclude that almost all of America's past positive or negative progress has been made possible by the unreasonable man (or woman). In recent years, America has been adapting to a small minority of unreasonable people who have, in my opinion, led to a downward economic and cultural, spiral. It's called complacency, or going along to get along. Let's turn this around by becoming unreasonable advocates for advancing the human condition and spreading God's Word.

EPILOGUE

This Epilogue is a continuation of my future predictions presented in the Prologue. But it's presented as an epilogue since many of the chapters in this book provide further background information that the reader needs to know in order to rationalize what is to follow. However, be forewarned, what I'm about to say is presented in unconventional descriptive language, and reflects my sense of logic based upon known science and my engineering background. It's comprised of unique thought experiments that to the best of my knowledge have never been posited before, and begins with the premise that God is the designer and creator of all things and a literal interpretation of the Bible.

In the Prologue, I describe a future that appears close to being utopia. Current concerns about overpopulation, climate change, water shortages, energy shortages, and an unequal distribution of wealth are nonexistent; however, that doesn't mean that the human tendency for evil has been eliminated. Although there is no reason for crimes such as stealing, there will still be emotionally driven crimes like murder. Although cancer, heart disease, and other life-threatening illnesses have been eliminated, there will still be birth defects and eventual death. Although there is no need for political differences, there will still be those who will want power and control, possibly leading to wars or even worse, nuclear annihilation. Although androids will be doing most of the required work, idleness removes the sense of self-worth, accomplishment, and pride that can only be derived from working for something that has purpose. Idleness has also been called "the Devils workshop."

With that being said, it appears that mortal humans will never achieve utopia because of evil-doing and the prospect of mortal death. On the other hand, with the world's population living in satellites in space, and having a universal communication capability, something might happen at that point in time that changes everything. That change might indeed bring about a real utopian existence. If you've been paying attention to what I've been saying throughout this book, I'll bet you can guess what that happening might be. You're right: it's at this point in our human existence that God may intervene. However, the Bible states that only those who qualify for immortality will be transformed. It's called "the Rapture." Those who are left behind will be given a second chance, but when the Rapture becomes history, it will have a profound effect on many of those who heretofore had ignored the handwriting on the wall.

So, – you may ask, where am I going with what I've just described? Ok, here it is. If you believe God is the designer and creator of all things, it's logical to assume He gave mortal humans information about their origin and guidance on how to live their lives. I believe the Bible contains that information and guidance and is literally true. However, many things in the Bible can be interpreted in ways that don't conform with mainstream theological doctrine. In fact, at the end of this Epilogue I will share what I believe is a ground-breaking discovery that might change the way theologians interpret the Genesis account of creation and what our mortal human purpose might be. However, before telling you what that is, I need to provide some additional background information

MANSIONS IN THE SKY

Imagine if Christ's parables were written in a clear and unmistakable language. If they were, they might not have made sense to the people who lived in the past, and probably not understood by most of today's population. Interpreting Christ's parables in light of today's technology may still not be sufficient -- especially when considering technologies that may exist hundreds (or even thousands) of years from now. Therefore, I wanted to interpret Christ's parables in my own way based upon current technology and my rationalized – prologue -- forecast of a future that involves living in space. In John, chapter 14, verse 2 Christ states, *"In my Father's house are many mansions: if it were not so, I would have told you. I go to prepare a place for you."* So what are these mansions? Are they Earth-like planets distributed throughout the universe that are ready for terraforming, or are they something else? If so, the wording suggests that these mansions may not have been prepared at the time of Christ's death on the cross. Maybe that's the job of future mortal (or immortal) humans with the help of advanced android-like robots. Whoa! You're probably thinking that that sounds a bit presumptuous. But let's continue.

In Mathew chapter 13 verses 31 and 32, when speaking about the kingdom of heaven Christ said, *"The Kingdom of Heaven is like a grain of mustard seed, which a man took and sowed in his field -- which indeed is the least of all seeds: but when it is grown, it is the greatest among herbs, and becomes a tree, so that the birds of the air come and lodge in its branches."* In this passage Christ is, of course, speaking in a parable. The mustard seed is so small that it takes more than seven hundred seeds to equal one gram, and if one pound equals 454 grams, it's pretty small. Yet, when the seed grows it becomes a tree of significant size. After hearing about this biblical passage in church one Sunday, it reminded me about how insignificantly small Planet Earth is relative to the size of the universe. This size relationship made me think Earth and its people might be like that mustard seed, and our human destiny might be to eventually develop and inhabit the vastness of the universe. The pastor's interpretation was that Christ's meager beginning has influenced billions of people throughout the world and will continue to influence more and more people. While this is true, my interpretation goes much further. However, with regard to my pastor's view, I believe America is the key to spreading the gospel to the world's growing population; and the pastor's interpretation of the parable is just a precursor to what is meant to follow.

Are Earth's humans meant to develop the technologies needed to develop other Earth-like planets, or to robotically manufacture entities for living in space? Are these Earth-like planets or entities the "mansions" that Christ spoke about in His parable? Open your mind and imagine what this could mean. To help you out, I suggest you refer to the Bible's description of the New Jerusalem referenced in Revelation (21:16 NCV). It describes a city (or space-based entity) that is *1,500 miles high, 1,500 miles wide, and 1,500 miles deep.* From what I can observe, God doesn't make large cubic structures; humans do! Generally speaking, God makes gravitationally formed objects like planets and moons. On the other hand, it's interesting to note that these 1,500-mile dimensions are close to the dimensions of our Moon.

Perhaps our human destiny is to create billions more planet Earths throughout the universe with the help of electronically powered, self-replicating robots. Could these Earths be populated by mortal humans in a way that differs from a primitive Adam and Eve scenario? Moreover, could our immortal souls, existing in human rather than spiritual from, be quartered in specially designed and robotically built dwellings in space?

Although God could create organic human-like robots programmed to build objects in space, their need for air, food, and other resources would make them impractical for working in an outer-space environment. Therefore, it makes sense to me that electronically powered, self-replicating, computer-programmed, humanoid robots could be mankind's greatest technological achievement. And, something for which God needs humans to accomplish in order to fulfill His plan.

ADAM AND EVE

Consider this, when God and His immortal angels, terraformed Planet Earth, He put Adam and Eve in the Garden of Eden. As beautiful as the Garden of Eden might have been, it didn't have a contemporary home with hot and cold running water and a dishwasher. So I think it's safe to say that the Garden of Eden was primitive, relative to today's standards. This then begs the question, why didn't God create robots and the blueprints needed to create a modern home and a Mercedes-Benz parked in the garage? If God intended the Garden of Eden to be so wonderful, why not go all the way? Does this mean that since God is a spiritual being, the technological needs and wants of human beings need to be created by humans? It's interesting to note, however, that God did provide all of the materials and the wherewithal for humans to invent and create a Mercedes-Benz (and other up-to-date conveniences). Are these human inventions and creations part of God's plan?

Since God is unquestionably the greatest inventor and creator of all time, it should be a minor task for Him to invent and create a Mercedes-Benz. On the other hand, maybe it's *not* possible for God to create a Mercedes-Benz. This might explain a lot of things related to our human purpose. Just because God can create life and transform atoms at His command (creating Adam from the dust of the ground), doesn't mean that these transformed atoms can be assembled into an automobile that requires a build-up of supporting technologies, manufacturing plants, an evolving knowledge base, spare parts, and a repair mechanic. In contrast, the things that God *has* created, baffle human understanding even to this day.

To make my point, might I digress for a moment? The last paragraph reminds me of a fictional story about a man who bought some scrubby land and worked hard to transform it into a beautiful garden. He planted trees and flowers in an artistic way. He dug a pond and used some of the leftover dirt, some large boulders, and an electric water pump to create a beautiful waterfall and stream leading to the pond. Creatively designed bridges were built to cross the stream and blend in with the background. Swirling walkways were designed to allow visitors to traverse through the garden and see the majestic views. Water fountains were added along with discrete signs that described each tree and plant. Benches, shelters, and restrooms were placed at strategic points to allow visitors a place to rest and be protected from an occasional rainfall. Needless to say, all of this was meticulously maintained in a pristine condition.

Okay. That's the mental picture I want you to have. Well, as the story goes, one day as a woman was passing through the garden, the man who created it approached the woman and asked, *"How do you like the garden?"* The woman replied, *"Isn't it wonderful what God has created?"* The man answered, *"You should have seen it when God was taking care of it!"*

Don't get me wrong. God has created some beautiful landscapes accompanied by spectacular sunsets, but humans can in some cases improve the view by using equipment and technology. We can use manmade equipment and devices like backhoes, concrete forms, and electric pumps for this purpose, while God creates the trees and flowers for man to arrange in creative ways.

MINING THE SKY

Will humans, with the help of self-replicating robots, ever be capable of terraforming? I say yes' it's just an engineering problem on a grand scale. Here's how, and perhaps why, we will eventually do this.

As you may recall in the Prologue, my prediction of the future included huge rotating satellites made by billions of self-replicating robots and specialized equipment for construction in space. The materials and energy were derived from asteroids and helium 3 contained in the upper atmosphere of the gaseous planets respectively. In case you don't understand what I mean, let me begin by telling you some things that might surprise you.

After reading John Lewis' book *Mining the Sky*, I found that the amount of helium 3 contained in the atmospheres of Uranus, Neptune, Saturn, and Jupiter is about 3,000,000,000,000,000,000 thousand pounds. When you consider 900 thousand pounds of helium 3 is enough fusion fuel to provide one year's worth of power to support the Earth's current population, you might begin to appreciate the potential of this power source and the need for mining in space: more than 3 trillion years of energy! But that's not all folks. Besides the vast amount of helium 3 fuel, the asteroids are estimated to contain the following vast amounts of other materials that could serve to support Earth's future population:

1,650,000,000,000,000,000 thousand pounds of iron;
115,500,000,000,000,000 thousand pounds of nickel;
8,250,000,000,000,000 thousand pounds of cobalt;
24,750,000,000,000 thousand pounds of platinum;

Plus, large quantities of other materials such as gold, silver, copper, manganese, titanium, rare earths, silicon, and uranium.

Putting these numbers into perspective, if we were to build a six-inch-thick hollow iron sphere to enclose the planet Mars, we would be using only 0.011 percent of the amount of iron available. However, as large you may think these numbers are, it's interesting to note that the total weight of all the known asteroids and comets only represent a miniscule fraction of Earth's total weight.

You might also be interested in one other piece of information. One asteroid in particular, Ceres, contains about five times as much fresh water as there is on Earth. For the Ceres asteroid, a freshwater upper limit was estimated to be 430,000,000,000,000,000 thousand pounds.

Now let's consider having our robots construct a hollow sphere that's about the size of our moon and making it a spacecraft. To simulate gravity let's make the equator portion a rotating hollow ring where gravity is simulated on the inside – outer – surface. Using a 1,500-mile high central column with propulsion capability, we then connect the ring and column with spokes. Now let's construct a geodesic dome on the top and bottom of the ring. If we now layer the outside of the resulting sphere with processed asteroid debris – left over from satellite construction – we will have something that looks like our moon, but with a propulsion capability and gyroscopic stability. Oh, I almost forgot. Let's bombard the surface with some asteroids to make it look like a naturally occurring object in space.

TERRAFORMING

Now you may ask, what does mining the sky, billions of self-replicating robots, and a moon sized spacecraft, have to do with terraforming? Well let's take the moon sized -like spacecraft on an interstellar journey to a preselected solar system that has a planet with earth formation potential. And while we are at it, let's follow it with a few thousand – Epsilon 1 type – spacecraft. Assuming the journey takes a hundred or more years, we grow an animal and plant population that will be used in the terraforming process. Because our moon spacecraft has a huge inner space we can develop compost to cover the surface of the selected planet. In addition, the chemical and/ or biological ingredients required for transforming a carbon dioxide atmosphere into a nitrogen/oxygen atmosphere could be taken on the trip to provide the planet with Earth-like conditions.

Upon arrival at the selected planet, the first order of business will be to place the planet in the right orbit and twenty-four-hour spin rate. This can be accomplished by using the moon sized spacecraft as a gravitational tugboat. Of course the spacecraft will then serve as the planet's moon to stabilize its spin axis and regulate tides.

How are we doing so far? Do you think this is fairy-tale thinking, or can you imagine a time in the future where all of what I'm saying is possible? Anyway I'll continue because there is an important point I want to make.

In general, the atmosphere of many naturally formed solid Earth-like planets is carbon dioxide and methane which might result in it being non-transparent. Thus when the selected planet's atmosphere is transformed, it will allow much more light to reach the surface. Coincidently, the Bible's Genesis account states that on the first day of terraforming planet Earth, God said, "Let there be light." So on the first day we place the planet in the right orbit, at the right spin rate and axial precession, and transform the atmosphere.

As a guide, I'll continue to refer to the Genesis account of Earth's terraforming process. Thus on the second day, we need to separate the land from the water. Here, I'm going to assume subterranean water and a drilling operation that releases a specified amount of subterranean water to the surface. On the third day we cover the land surface with compost that developed over the hundred-year travel time, and plant grass, trees, and other forms of vegetation.

On the fourth day of the Genesis account, God created the "lights in the sky" – the sun, moon, and stars. In our terraforming project, these heavenly bodies already exist (I'll explain more about this later), so let's proceed to the fifth day, which includes transporting animals, birds, and fish to the planet's surface and waters. Then on the sixth day God created Man. Since, in our case men and women already exist, we might ask for volunteers to live on the planet and move on to the next step, modernization.

So what is modernization and why didn't God do this? Well as you know by now, I believe that our Earth was terraformed by God and it took thousands of years for its human inhabitants to finally modernize it. Whether or not God did the same thing in other solar systems is certainly unknown.

Now you may ask, how did God terraform planet Earth without humans and robots? You may also be asking why mortal or immortal humans are going to do the terraforming in the future if God can do it. These are logical questions and here's my answer.

If you recall at the beginning of this book, I speculated that God most likely created visible matter (atomic structures) with command and control capability. Thus explaining how Adam was created from dust. If true, it's not inconceivable that a four-billion-year-old planet, moon, and Jupiter combination, with the right size and density (composition) and located in the right orbit in the right solar system, and in the right portion of a spiral disk galaxy, could be located and terraformed by God. However, as previously stated, the number of planets with these right conditions would be extremely rare and adjustments would be needed to create the correct starting conditions needed for terraforming. So, you may ask, why couldn't God do this? My answer is that God may be able to transform atomic and molecular structures, but moving randomly placed planets and moons into optimum locations requires more than matter transformation. In other words, when God created the universe, probably from dark matter, it was probably done in a somewhat random fashion, and required a search for the right existing starting conditions for terraforming Planet Earth.

Also, as you may recall, I speculated that when Adam and Eve were created in a Garden of Eden it was primitive relative to today's standards, and it's taken more than six thousand years for mankind to modernize it. Moreover, most of today's modernization has happened in just the last two hundred years, a condition I believe explains a lot about one of the purposes of mortal humans: the development of space technology and self-replicating robots.

IN THE BEGINNING

Based upon what I've just said, here is another thought provoking experiment. In the beginning, perhaps at zero time billions of years ago, there was God. God permeated the universe which *"in the beginning"* was dark matter. At first God created galaxies and solar systems, but the stars had not yet been ignited. God then located a watery (icy) planet (without form) and moon combo that was ready for terraforming. God then said, "Let there be light," and creates the planet's transparent atmosphere, which is then bathed in the light of God. Since God can exist in zero time, a day could be any length since at this point a day had not yet been invented. So for the next and following time periods "days" a planet is terraformed by atomic and molecular transformation, however in the fourth time period (day) God ignites the suns in the galaxies, which appear as a sun, moon and stars to an observer on the planet (God adjusts the speed of light to make this happen). Then during the sixth time period, God creates male and female humans.

I need to stop at this point and say that this terraformed planet may or may not have been our Earth, as you will see as we proceed. In fact, this may have been God's first creation of mortal humans who would later become immortal angels/ or "sons of God." Don't panic and disregard what I'm saying until you read further. In this thought-provoking scenario, these mortal humans, and their offspring, have a free will that's sometimes prone to evil-doing, but God intervenes to speed up their technological progress. Thus the process of technological advancement is accomplished in a very short period of time; a time in which self-replicating robots are created which are capable of creating large objects in space and terraforming planets. As a result, perhaps all of the then existing human population is given immortality and their population growth is stopped.

At a later point in time our Planet Earth is terraformed by the immortal beings and Adam and Eve are created. Six thousand years later, the offspring of Adam and Eve reach a point where technology allows them to create self-replicating robots and live in space. After living in space and developing terraforming technology God finally takes, raptures, those who qualify for immortality and condemns the remaining humans to a period of reform, a second chance.

Before going further, I assume the reader is familiar with the words "sons of God," which are mentioned in the Genesis account of creation. If not, look it up and consider what I'm saying about immortal beings being created before Adam and Eve. Wow. This is certainly out-of-the-box thinking, but ask yourself, have I violated anything that's said in the Old or New Testament, including prophesy? Let's see,

Based on the above thought experiment, it could be concluded that God requires immortal beings to first experience human mortality. In so doing, He demonstrates the chaos that results from a free will that's capable of evil doing, and why it needs to be voluntarily controlled when given immortality. Then, even when a mortal human qualifies for immortality, there appears to be a need for further evolvement through a hierarchy process. A process that may begin with being a watcher which is then followed by becoming an archon, aeon, archangel, and possibly some other steps. Don't be alarmed at what I'm saying so far; remember, this is just a thought experiment.

With regard to those who do become immortal beings in heaven, the wonders of the universe are probably revealed. However, their first assignment might be to watch over mortals and assure that the mortals don't destroy themselves by overpopulation (starvation) or nuclear annihilation. Depending on the outcome of each planet's inhabitants, and the management capabilities of the immortal watchers, it might be possible for the watchers to be promoted to where they serve directly with God by eventually becoming archangels. In our Earth's case, however, it appears that the male watchers (immortals from God's first terraformed planet) didn't do a very good job. By coming to Earth and mating with Earth women, their offspring contaminated Earth's population and God wiped them out in a great flood and started over. Further incompetence on the part of the Earth's watchers probably became evident to God in the very slow technological progress that followed the flood. Perhaps immortal beings have in recent times helped George Washington, Werner Von Braun, Albert Einstein, Nicola Tesla, and others to assure that America would come into existence and get technological ball rolling.

THE IMMORTAL SONS OF GOD

Obviously, what I've said so far has stirred up a hornet's nest in the minds of many readers. But since many readers may be unfamiliar with the Bible's mention of the "sons of God," I'll briefly divert from my terraforming thought experiment and explain. Besides it's a subject that interests me as a person who wants to know that everything in the Bible is literally true and makes sense.

To most Christians there is only one Son of God. However, the Bible does state that there were interactions between humans and the *sons of God*. In Genesis chapter 6 verse 2, *"The sons of God saw the daughters of men that they were fair; and they took them wives of all which they chose."* And, in Genesis Chapter 6 Verse 4, *"There were giants in those days; and also after that, when the sons of God came in unto the daughters of men, and bare children to them, the same became mighty men which were of old, men of renown."*

In case you're thinking the *sons of God* are just extraterrestrials that evolved on another planet, I'll say this. Darwinian-style evolution is absolutely not possible! If it was possible, your greatest of great-grandparents were miraculously formed single-cell bacteria, not apes or chimpanzees. It just doesn't make logical sense, regardless of billions of years.

Lewels' Book

Being curious about the immortal *sons of God,* I found a book entitled *Rulers of the Earth* by Joe Lewels, PhD In this book the subject of the *sons of God,* and the origins of the Bible are discussed quite extensively, but the book is very controversial to say the least. While I don't subscribe to many of the Lewels' conclusions he has done a good job in researching the subject of the *sons of God,* and the origins of the Bible. Here is a brief summary.

In 1947, the Dead Sea Scrolls were discovered in some clay jars in caves located in a place called Qumran. After forty years of deciphering the contents of the scrolls by the Catholic Church they were finally made public. Once sorted out, they can be categorized into four types of books. The Torah (the first five books of the Old Testament), books of commentary on the scriptures, books of secret occult wisdom, and books on astrology. Included in the books of secret occult wisdom is the book of Enoch. Included in the book of Enoch, is information about the angelic hierarchy that oversees life on Earth. The angelic hierarchy is described as watchers, archons, aeons, and archangels. Also included is a history of the world before the great Flood, when angelic beings descended from the heavens and mated with humans.

Certainly if there was a flood and one man and his family survived, it would make sense that that family would know about the hybrid beings and their parentage. They would have been there and seen what was happening firsthand. Also, since Enoch was the great-grandfather of Noah, it would make further sense to believe that Noah had possession of, or knowledge of, the book of Enoch.

Could construction of the Tower of Babel, shortly after the Flood, plus other pyramid structures around the world, be testimony to a belief in trying to reach the sky in an effort to communicate with heavenly beings? With Noah and his family having firsthand knowledge of these beings, it might explain why people of that time would expend so much energy to create these structures.

As an aside, for those who are not familiar with the Bible, or the Tower of Babel, it was a time when a tower was being constructed and the language was "confused," apparently by the *sons of God.* Genesis chapter 11 verse 7 states, *"Let us go down there confound their language, that they may not understand one another's speech."* The reason for doing this is not clear. However, because the continents had separated during the Flood, and land bridges could have existed at that time (re: my commentary about the Flood). It might have been a way to scatter the population to all parts of the Earth. When they did scatter, they would have carried with them a common knowledge and belief. Does this explain how the different languages, the common stories about the Flood, a belief in gods mating with humans, and similar worldwide pyramid construction came into being?

Mention of the *sons of God* in the Bible's Old Testament is somewhat unusual when you consider that it seems to fly in the face of the notion that there is only one son of God. If the *sons of God* were immortal angels, how could they have human bodies and mate with mortal women? It's even more unusual when you consider that at least one word in the Torah has been revised by some unknown editor. According to the Dead Sea Scroll version of the Torah, Deuteronomy chapter 32 verse 8 uses the words *sons of God* rather than the current biblical phrase, *sons of Israel.* If this was changed why wasn't *sons of God* revised in the book of Genesis? In any event, the words in Genesis have survived to this day, in spite of the apparent confusion that it causes for the Christian and Jewish religions. My point here is not to cast doubt upon the Torah. Instead, the fact the author Moses left the words *sons of God* in the Torah, adds to its credibility.

Other ancient writings and religious beliefs around the world also support the idea of a common ancestry and the interbreeding with gods. The ancient Sumerian Kings, the pharaohs of Egypt, the Incan rulers of Peru, and the dynastic kings of China and Japan claim to have been the offspring of gods. If these rulers were the offspring of the *sons of God,* then the biblical account of the Flood would be in doubt because these offspring should have been destroyed during the Flood event. However, a common heritage and a belief by their followers in the *sons of God*, would explain why these leaders would claim a divine right to lead. It may also explain a basis for the Egyptians, Greeks, and Romans to believe in multiple gods.

The more I investigate ancient cultures and writings, the more I find a common thread, that indicates a common ancestry for humankind. Of course the most pervasive writings involve the flood, but as stated above, the common belief in the offspring of gods can be traced back to a common starting point. Just keep in mind, that no matter how outlandish some of the ancient writings may sound to our current thinking, there is always an element of truth that probably should not be ignored.

Schroeder's Book

Some theologians are buying in to the evolution scenario, that the six days of creation are actually billions of years of cosmic time, as described in Dr. Schroeder's book, *The Science of God*. However, they believe that God, rather than natural processes, terraformed Earth and created life. From this perspective, they believe that Adam, Eve, and their male offspring were the *sons of God*, with souls, and the *"daughters of men"* were pre-human beings without souls, perhaps Neanderthals, thus making the argument that Adam and Eve were Neanderthals who were given souls by God. This would explain why the soulless hybrid offspring of Neanderthals displeased God to the extent that He created a flood to wipe them out.

My problem with this possibility is that it doesn't read that way in the Genesis account. Genesis chapter 2 verse 7 states, *"And God formed man of the dust of the ground, and breathed into his nostrils the breath of life; and man became a living soul."* It doesn't say that God made a living soul from a Neanderthal.

So there you have it. In one case you have God creating humans from Neanderthals that were a result of millions of years of Cambrian-derived macro-evolution. Conversely, we have Adam being created by God from the atoms contained in the dust of the ground. Of the two possibilities, I believe the latter case makes much more sense.

While examining other explanations for the *sons of God,* I've looked at other possibilities to see if any of them made sense. Of these possibilities I found one that's worth mentioning because it's based upon intriguing non-biblical ancient writings, and it's called *The Epic of Gilgamesh.*

The Epic of Gilgamesh

The story of Gilgamesh was written by the ancient Sumerians. It tells of a man who survived a worldwide flood which scientists claim predates the Bible's story of Noah. Without going into a lot of detail about the Gilgamesh story, I recommend that you check it out on the internet. In summary, the Gilgamesh story is about a superhuman (one third human and two thirds god) ruler of a Sumerian city, who was trying to achieve immortality. To achieve immortality, Gilgamesh befriends a human who had been given immortality by the gods. However, the gods had collaborated to destroy the Earth in a flood, apparently because the gods couldn't sleep! However, one of the gods alerted Gilgamesh's friend to build a boat as a way to escape the consequences of the flood. When the flood came, the boat saved Gilgamesh. However, in the end, Gilgamesh did not achieve immortality.

When you read the Gilgamesh story, interpreted from twelve large stone tablets, the oldest presumed written account in history, you will find that much of it is written like a child's fairy tale. The gods behave much like humans in that they were fighting, plotting, and deceiving each other; and the 200 x 200 x 200-foot cubic-shaped ark was built in 7 days, whereas Noah's 450 x 75 x 45-foot ark was built in 120 years. In the Noah story, God destroys life on Earth because of the evil created by the hybrid children derived from the immortal *sons of God*. In the Sumerian story, the hybrid child, Gilgamesh, is saved. To compare the two stories I recommend that you read the account written by Rich Deem, , which concludes as follows:

> *'We have examined the similarities between the Epic of Gilgamesh and the Genesis flood account of the Bible. Although there are a number of superficial similarities between the accounts, the vast majority of similarities would be expected to be found in any ancient flood account. Only two similarities stand out as being unique – landing of the boats on a high mountain and the use of birds to determine when the flood subsided. However, both of these similarities differ in important details. In addition, there are great differences in the timing of each of the flood accounts and the nature of the vessels. Why these details would be so drastically changed is a problem for those who claim that the Genesis flood was derived from the Epic of Gilgamesh."*

On the other hand, I see something very peculiar about the Gilgamesh story, in spite of the fairy tale aspects that might account for the survival of at least one hybrid human after the flood. Could the Gilgamesh story be a separate account describing the behavior of the immortal *sons of God* before the flood? Could there have been another ark built when Noah was building his ark? Since it took Noah 120 years to build his ark, wouldn't the immortal *sons of God* have seen what was going on and figured out what God had planned for the destruction of their hybrid children? Moreover, if the story of Gilgamesh is the oldest story ever written by humans, why would it be a fairy tale? Wouldn't the writers want to say something significant about what actually happened, especially if they took the time and effort to carve it into twelve large stone tablets?

A PROVOKING THOUGHT

Now getting back to my terraforming thought experiment, here's my conclusion to all of what I've said in this book. While what I'm about to disclose may appear to go against mainstream Judeo-Christian teaching, I can assure you that nothing I'm going to say is anti-biblical. In fact, it is a literal interpretation of the Bible from a modern-day perspective. Because our technology has moved so rapidly in recent years, I believe mainstream theology needs to catch up with what God is currently revealing. The advent of electricity, television, remote controls, lasers, computers, robotics, satellites, space exploration, particle accelerators, nuclear fusion, global positioning, and the dimension of time, changes almost everything we thought we once knew.

When we consider God beginning the universe, wouldn't He want to create immortal beings with a free will? If so, wouldn't He want to give them a human and spiritual form so that an eternal life would include endless unimaginable pleasure and purpose? If true, perhaps the first order of business might have been to create mortal humans that develop technology that's useful to mortal humans. To make this happen, perhaps there needed to be an Earth-like planet for these mortal beings to eventually create technologies; thus, my speculation that an Earth-like planet was created before our Planet Earth, with the mortals from this planet becoming immortal *sons of God,* who later terraformed our planet. Because these immortal beings had a free will they were capable of evil doing. Perhaps God then realized that He needed to correct this human "free will" deficiency by establishing a selection criteria and an incentive plan that rewarded immortals who were mature enough to suppress their evil tendencies.

So when the Bible says, *"In the beginning God created the heaven and the earth,"* it's possible that this earth was created at a different time and place in the universe. The Bible then goes on to say on the sixth day, *"Let us make man in our image after our likeness."* Prior to this point the words *"let us"* were not used. Therefore, when and if God's immortals (from the first Earth-like planet) terraformed our Planet Earth, He was working in conjunction with them at this new point in time (the sixth day). If the Bible had said, *"In the beginning God created heaven and the first Earth and the sons of God,"* and then went on to say *"Then the sons of God created a second Earth,"* this would be more than confusing. However, by using the words *"let us"* on the sixth day of our planet's terraforming process, God is clearly saying there were other beings with human characteristics created before Adam and Eve. By going on to describe the *sons of God* interacting with mortal human women, the message is also clear that the *sons of God* were every bit as human as we are, but immortal. In other words, they look like us (in our image) and can eat, drink, and were in all other ways human!

Let me describe the last paragraph in another way. When the words *"In the beginning God created the heaven and the earth"* appear in the Bible, I would say this is God's universe-creating big bang. However, it's clear to me that after the big bang God created immortal humans that are described as the *sons of God,* in the Bible, and that these *sons of God* have human characteristics in that they are capable of mating with mortal human women. If they have human characteristics, it's logical to assume they would need air to breath and food to eat. Therefore, it's also obvious to me that they need an Earth-like environment.

Since it's illogical to assume that God made space ships with breathable air, the Earth-like environment must have been a terraformed planet. The terraformed planet could have been a planet before our Earth or Earth itself. To me, both of these hypotheses are possible. Thus, if our Earth was the first and only Earth-like planet in the universe, there would have been a time interval between the fifth and sixth days of creation that allowed the *sons of God* to come into being and to develop technologies that included living in space. So on the fifth day we need to include the creation of soon to become immortal beings, as well as animals, fish, and birds. We also need enough time for these first mortals and their offspring to develop space technology and be given immortality. Since the Bible doesn't explicitly say that immortal humans existed prior to Adam and Eve, it implies it when God says *"Let us make man in our image after our likeness."* In this case, it appears that God is speaking to previously created immortals.

Moreover, if our Earth was the first and only one created (terraformed), and added the time needed to develop advanced human technology, it would imply a significant number of Earth years and a significant growth in human population. This makes sense when the words *"let us"* are used. Because of the words *"let us,"* it's logical to assume that either there were not very many immortal beings at that time, or God was speaking to only a few immortal leaders (archangels). It also makes sense that all of today's human technology was developed prior to the existence of Adam and Eve. It makes sense because God probably needed to have a test case (engineers call it making a prototype before going into production) to insure Earth's terraformed conditions facilitated His planned result for mortal humans—space technology, self-replicating robots, and immortality. This then begs the question, why did God create mortal humans if the space technology had already been created? To me the answer is obvious. God didn't want to create more immortal beings like the first group who had not matured enough to rid themselves of evil-doing and were capable of turning against Him. There needed to be a progression of steps that suppresses the free will from evil-doing, and mortal beings are intended to be the beginning of this process.

Since some of the first immortal beings, *sons of God,* were probably given the responsibility of watching over us, we have probably on various occasions encountered them with UFO sightings and other instances. Perhaps they have left evidence of their encounters like crop circles, pyramids, movement of large stone objects, and other unexplained findings that might be part of their pre- and post-Flood legacy. After all, many of these immortals would have had a dubious past since they were probably given immortality before God apparently changed the selection and development criteria. That might account for at least one high-ranking immortal opposing God. *The archangel named Lucifer comes to mind.*

GOD'S SON

Finally, I want to end this book with how Jesus the Christ fits into all of what I've just said. The Bible states that God sent His *"only"* begotten Son to planet Earth to provide an escape to immortality for any mortal human through simply acknowledging and believing that He is the *"direct offspring"* Son of God. To me, this means that by verbally expressing your intent to follow Christ's leadership in your life you will avoid continuous mortal rebirth (again just my current opinion) and become at least a first-level immortal being when you die or are taken away while you are alive when Christ returns. And, from all that I can determine, an immortal life means that you will experience wonders beyond imagination.

By sending His son to become human and die a horrific death for our evil doings and anti-God behavior, God is showing that a penalty must be paid to assure justice. He is also demonstrating love for His children – all of mankind – by providing a pathway to freedom from the bonds of mortal pain, anguish, and death.

A WRAP-UP

When Charles Darwin wrote his book *On the Origin of Species* in 1859, it was widely rejected by the scientific community at that time. However, by the 1870's it was widely accepted, and for the most part, it still is a commonly held belief. Because of work done in the 1960's, by Noble Prize winners Francis Crick and James Watson, we now know that the DNA molecule contains specific information – similar to that found in a computer code – that directs the actions of twenty different types of proteins, that combine to form a living self- replicating cell. Moreover, the proteins themselves act like a computer that processes the DNA's coded information. As a result of this finding I am justified in proclaiming that we now have "undeniable" proof of intelligent design. This finding – by itself -- totally disproves the Darwinian theory, and it's just a matter of time for the general public to recognize this fact.

Since intelligent design is a statistically proven fact (equivalent to more than all the atoms in the universe), the only recourse left to God rejecting scientists, is to proclaim – out of desperation -- an infinite number of universes. However, based upon common sense, observation, and the many other intelligent design related facts presented in this book, it should be "self-evident" to any thinking person, that God exists; and that He is the creator of the universe and all things therein -- including us. With this thought in mind, I believe we humans should begin looking for further evidence that might reveal additional information about God's cosmic plan for humanity. In this regard, I've looked for clues that God may be providing with our modern day scientific discoveries.

About two thousand years ago God sent his Son to reveal – among other things -- that there is life after death. With our recent fast moving scientific discoveries, I believe God is again trying to reveal additional information about his afterlife plan. While the terraforming thought experiment provided in this epilogue may – or may not – be correct, it should be considered as a starting point for others to consider and/or provide alternative scenarios. After all – except for mansions in heaven -- the Bible is not very clear about what we will experience in our eternal afterlife.

MY BACKGROUND

My degree is in mechanical engineering from Northeastern University in Boston, Massachusetts. Upon graduation in 1964, I was commissioned a lieutenant in the United States Army Corps of Engineers and served in Vietnam.

In August 1966, I began my career with the Westinghouse Electric Corporation. During the first seventeen years, my experience was primarily in nuclear steam generator development. During the last four years of these seventeen years, I was promoted to Section Engineering Manager and, later, Department Engineering Manager of a $40 million Atomic Energy Commission (presently the Department of Energy (DOE)) contract, to develop a steam generator for a "breeder" nuclear power plant. My department also performed government contract engineering studies to develop solar powered steam generators and fluidized bed coal gasification boilers.

In 1982, I transferred to the Thermo King Corporation (a subsidiary of Westinghouse) to become Engineering Manager for the Truck Transport Refrigeration Equipment Department. During the early portion of my employment with Thermo King, my engineering group developed a new line of truck refrigeration equipment for the world market. I was later given responsibility for design engineering activities at Thermo King factories in Barcelona, Spain; Hamble, England; and Prague,

Czech Republic. Before retirement in 2001, I was responsible for the successful development of a new line of refrigeration equipment for the Japanese market and the development of a new alternator-powered refrigeration system for the European market.

I have written numerous technical papers for international conferences and for the American Society of Mechanical Engineers (ASME) and currently hold seven patents. I was also the chairman of the Florida West Coast Section of ASME in 1974–75 and passed the state of Florida professional engineering examination in 1975.

CPSIA information can be obtained
at www.ICGtesting.com
Printed in the USA
BVHW07s0927090918
526908BV00001B/26/P